Hands-On ZigBee:

Implementing 802.15.4 with Microcontrollers

Hands-On ZigBee:

Implementing 802.15.4 with Microcontrollers

By

Fred Eady

AMSTERDAM • BOSTON • HEIDELBERG • LONDON
NEW YORK • OXFORD • PARIS • SAN DIEGO
SAN FRANCISCO • SINGAPORE • SYDNEY • TOKYO

Newnes is an imprint of Elsevier

Newnes is an imprint of Elsevier
30 Corporate Drive, Suite 400, Burlington, MA 01803, USA
Linacre House, Jordan Hill, Oxford OX2 8DP, UK

 Recognizing the importance of preserving what has been written,
Elsevier prints its books on acid-free paper whenever possible.

Library of Congress Cataloging-in-Publication Data

(Application submitted.)

British Library Cataloguing-in-Publication Data
A catalogue record for this book is available from the British Library.

ISBN-13: 978-0-12-370887-8
ISBN-10: 0-1237-0887-7

For information on all Newnes publications
visit our Web site at www.books.elsevier.com

07 08 09 10   10 9 8 7 6 5 4 3 2 1

Contents

Contents

Preface

My friend Jim caught word that I was writing a ZigBee-related book. Knowing that I have only written about technical things in the past, Jim asked why I was writing about an obscure 1930's magician. I've never heard of the Great ZigBee but Jim must have come across him somewhere in his travels.

This book is not a collection of magic tricks and illusions. (Although sometimes I think that RF engineering is an illusion. Think about it. Why do you have to shield RF stuff? Maybe because something very, very evil is going on inside the box? Hmmm…) This book is all about teaching you about the IEEE 802.15.4 specification and how you can apply it to your own projects using IEEE 802.15.4-compliant development tools and radios. As you will find out, IEEE 802.15.4 and ZigBee are not one and the same. So, as we discuss the nuts and bolts of IEEE 802.15.4, we'll also discuss the components that comprise ZigBee.

Not too many computing gadgets are restricted by wires these days. Just look at the do-it-all wireless camera-toting, spreadsheet-running, text-messaging, internet-browsing, emailing platform we call a cell phone. The do-it-all-with-RF philosophy of today's cell phones has spread to the ISM (Industrial, Scientific and Medical) sector. Sensors, test equipment and medical instruments are cutting the wires and replacing them with small short-range networks that require little human intervention to operate. In most cases, these small networks are based on the IEEE 802.15.4 specification. There are an unlimited number of additional applications for small IEEE 802.15.4 networks and that's why I believe IEEE 802.15.4 networks and ZigBee are working solutions that engineers will adapt into our everyday lives for many years to come.

As I write these words, the ZigBee Alliance is tweaking their ZigBee specification. That's OK. Change is sometimes good. On the IEEE 802.15.4 side, those guys and gals are in the Bahamas. Things ain't gonna change too quickly over there. What that means is that the backbone of a ZigBee network, IEEE 802.15.4 networking, won't be changing significantly even though ZigBee may be going through a phase of expansion and change. Rest assured that when you finish this book you will know enough about IEEE 802.15.4 and ZigBee to make intelligent design decisions, no matter what the ZigBee protocol finally turns out to be.

While we're on the subject of protocols, many of you have asked when my 802.11g book will be available. Well, don't hold your breath. I had to pull hen's teeth to put the 802.11b book together, as no one in the industry would step forward to help me with the technical details of the 802.11b radios. I've sent messages and placed calls to the folks that I thought would have an interest in providing you with the how-to's of embedding an 802.11g card into a microcontroller-based platform. I've received nothing in return. The same goes for SDIO. SDIO folks won't even return my calls. So, scratch both of those subjects off your wish list for now.

I can't complain about the ZigBee and IEEE 802.15.4 crowd. Out of all of the requests for assistance I generated, only a couple of them were denied or ignored. ZigBee Alliance members even referred me to other members of the ZigBee Alliance for answers and product assistance. With that, I would like to thank every IEEE 802.15.4/ZigBee product vendor and ZigBee Alliance member that returned my phone calls, answered my emails, put up with my whining and contributed to the content of this book.

You should understand that there are some things you will not get out of reading this book. This book is not a ZigBee stack doctorate-level dissertation and you won't become a ZigBee stack expert by reading this book.

Most of the IEEE 802.15.4 radio IC manufacturers offer a complete set of Gerber files that will enable you to clone their radio module if that's your goal. Building an IEEE 802.15.4-compliant radio from scratch is not something you and I will explore in this book.

If you are not familiar with microcontrollers and the communications protocols they employ, there are no in-depth tutorial discussions involving SPI and RS-232 in the pages of this book. I also pretend (we will never assume anything in this book, as assumption can turn you into a donkey) that you know enough basic C to follow along with the code examples I'll be offering to you throughout the book. If you're challenged in these areas, there are many fine Elsevier titles that you may find helpful.

On the other hand, there are many things that you *will* take away from reading these pages. The ZigBee protocol is based on the transport mechanism provided by an IEEE 802.15.4 network. You and I will begin at the lowest layers of the ZigBee stack, the IEEE layers, and work our way up. I will place a strong emphasis on helping you to understand IEEE 802.15.4 networking, as you will find out that for many of your applications ZigBee network functionality provided by the ZigBee protocol is not a prerequisite for passing meaningful data between network nodes. In fact, for many applications, ZigBee networking functionality is overkill.

All of the development kit hardware in this book is the latest and greatest. My goal is to expose you to as many of the popular IEEE 802.15.4-compliant/ZigBee-ready devices available to you. So, you'll get a good dose of IEEE 802.15.4-compliant/ZigBee-ready hardware along the way up the ZigBee stack.

Schematic diagrams of the various IEEE 802.15.4-compliant/ZigBee-ready radios and development kits will be scarce in the paper pages of this book. It is not my intent to reprint datasheet information that you can download from the manufacturers' web sites. However, if there is any supporting documentation that is important to the idea I'm trying to convey to you and it's not something you can easily download, you will find it on the Companion website that accompanies this book.

My wife says I talk too much and my Mom says that, when I do talk, nobody knows what I'm talking about. So, with that I'll shut up and let you get on with climbing through the layers of the ZigBee stack. I sincerely hope you have as much fun reading this book as I have had writing it ☺

What's on the Companion Website?

Following is a list of items included on the accompanying Companion website:

- UDP (User Datagram Protocol) Primer
- ARP (Address Resolution Protocol) Primer
- Frame Thrower II Source Code
- Rabbit 3000 Hardware Reference
- Cirronet Schematics
- Texas Instruments CC2420 Project with source code
- Microchip MRF24J40 Project with source code

Speaking the Language

Some famous person once said, "No matter where you go, there you are." No matter where you may find yourself, it's always a good idea to know how to speak the language. For instance, when you find yourself in a foreign country, phrases like "Thank you," "Hello" and "I need to use your bathroom" can come in very handy, depending on the situation. The same philosophy holds true when it comes to learning about things technical. To that end, I'm going to take the stance that you are reading this book because you want to know more about IEEE 802.15.4 and ZigBee. In this chapter, I'm going out one step further and will work with the pretense that you know absolutely nothing about IEEE 802.15.4 or ZigBee. So, in my mind, my job is to teach you to speak the language of ZigBee and subsequently IEEE 802.15.4. Once you become familiar with acronyms like PSDU, MPDU, NWK, PAN and MAC, the rest of the ZigBee puzzle and the underlying IEEE 802.15.4 protocol will begin to come together more easily for you. You'll need to understand the nuances of ZigBee-speak and IEEE 802.15.4-speak before we move into the actual hands-on portion of this book. Otherwise, you and I will just be mindlessly assembling meaningless ZigBee radio building blocks and crawling through undecipherable C source code.

If you're new to the concept of ZigBee, I want to introduce you to the world of ZigBee the right way. So, we'll start at the beginning of ZigBee time. However, first things first. It is required that you carve the ZigBee international anthem into your memory bank. So, get up out of that recliner and repeat after me:

> ZigBee: Wireless control that simply works.

> *Say it again, louder:*

> **ZigBee: Wireless control that simply works.**

> *I can't hear you!*

> **ZigBee: Wireless control that simply works.**

A True Story about a Couple of Flying Bugs

I don't know about you, but for me the ZigBee pledge of allegiance sets the tone within which ZigBee was initially designed. However, ZigBee wasn't always ZigBee, if you know what I mean. I actually came across an old Philips presentation dated June 26, 2000 that laid out the plans for what was then called the Philips RF-Lite Program. Interestingly enough, the presentation begins with a painting called *The Tower of Babel* and in the liner notes there are associated intellectual comments and interpretations of the Tower of Babel story. Although I found this to be pretty weird, of course, the idea was to convey that RF-Lite would cut through the clutter of remote control and existing competitive radio systems of the day.

As you progress through the Philips RF-Lite presentation, the concept of Firefly is presented—yet another protocol, but named after a flying bug. It is strongly believed that ZigBee is supposedly named for bees doing that zig-zag thing they do that always seems to find the best flowers and the hive. So, get used to references to flying bugs, particularly the bee.

By the way, ever wonder where lightning bugs (that's what us Southern kids called fireflies) went during the day?? Well, let me tell you. Those little buggers hide amongst the trees and vegetation during the day. And, if you think fireflies are docile little creatures that glow for our delight, get real. These guys come out of the gate eating meat and that glow they produce is for finding sex in the city. It has been said that if a male flashes the wrong signal, a female of another species of firefly will descend upon him and, yep, you guessed it, eat him up.)

Firefly was a spin-off of HomeRF-Lite, which was touted as a very low-cost method of low-speed data transfer that consumed very little power. (Hmmm…keep that low-power definition of HomeRF-Lite and Firefly in your long-term memory cells, as I guarantee you'll experience a little déjà vu as you continue to learn more about ZigBee and IEEE 802.15.4 networking.) A Firefly working group was formed and face-to-face quarterly meetings were planned. (Hmmmm…as I write this text, a "quarterly" ZigBee Alliance Open House is being held in Seoul, Korea and a new tactical direction for ZigBee was announced.) The Firefly working group included some big guns such as Panasonic, Texas Instruments, Honeywell, Invensys, Lego and Mattel. (Hmmm…Can you say Chipcon…Can you say Texas Instruments…Can you say Chipcon IEEE 802.15.4 radios now owned by Texas Instruments? You'll also see Panasonic play a part in IEEE 802.15.4 and ZigBee if you continue to read this little ZigBee book. Honeywell and Invensys are currently ZigBee Alliance members along with Texas Instruments. Hmmmm…)

Firefly was to be a low-speed product with a data throughput minimum of 10 Kbps and a maximum throughput of 115.2 Kbps. I can't provide any hardcopy proof that the Firefly communication link speeds were chosen to interface to and/or replace existing products with existing RS-232 serial ports, but those Firefly speeds fit nicely within practical minimum and maximum speeds of a typical RS-232 serial port.

The projected range of Firefly-based products was specified between 10 and 75 meters. Node count for a Firefly network maxed out at 254 with a maximum of four of what was termed "critical" devices. Up to 100 Firefly networks could be co-located. The real kicker was that a Firefly node would cost less than $3.00 and be able to operate up to 2 years before the Firefly's tail light would go dark.

The license-free ISM band was chosen for Firefly nodes, which would automatically insert and disassociate themselves in the network as required. Firefly was obviously aimed at indoor/outdoor use, as the marketing requirements listed RF penetration through walls and ceilings and home/garden use. To help eliminate interference with other devices operating in the same

ISM bandwidth, the Firefly nodes would use DSSS (Direct Sequence Spread Spectrum) technology and Carrier Sense Multiple Access (CSMA-CA) listen-before-transmit network access.

The Firefly protocol stack looked much like stacks look today. Data from the Firefly radio entered via a PHY (Physical) layer. The Firefly MAC (Medium Access Control) layer accepted data from the PHY layer and pushed it up to the Firefly DLC (Data Link Control) layer. The application layer resided inside the Firefly node and was accessed by an external user interface via API (Application Program Interface) calls. The configuration I just described would have resided in a Firefly slave node. A master Firefly node inserted a NWK (Network) layer between the DLC layer and the application layer.

As you've already ascertained, a master/slave relationship formed the Firefly network topology, with the master node being in direct communication contact with each slave node. This type of topology creates a virtual peer-to-peer communications link as every node can talk to every other node in the star as long as the master can pass the messages between the slave nodes that need to talk to each other.

Meanwhile, there were a number of engineers that had determined that Wi-Fi (the subject of my previous book) and Bluetooth, which is now trying to come into its own, as everyone has a Bluetooth cell phone interface in their ear, weren't going to cut it in some applications. What these guys and gals wanted was a self-healing ad hoc network of digital radios that could organize themselves into a cohesive and orderly network without external intervention. Simply put, they wanted a working Firefly. However, that was not going to be. The IEEE was already working on 802.15.4 by this time and the IEEE 802.15.4 standard came to life in 2003 and learned to fly (not glow and fly) in 2004. While all of this IEEE stuff was going on, the ZigBee Alliance sprouted from the ground in 2002 and it now looks like the ZigBee acorn is going to be a really big tree. As for the Firefly project, we all know what happens when fireflies are left in the jar too long.

Déjà vu

Although Firefly's tail light was glowing more dimly in the jar as each hour passed, many of the attributes of Firefly would find their way out of that jar and into ZigBee. Remember this? Firefly was a spin-off of HomeRF-Lite, which was touted as a very low-cost method of low-speed data transfer that consumed very little power. Well, how about this? ZigBee was designed as a low-speed means of data transportation, which consumes very little power and can operate for months or even years on a single battery. When compared with other wireless communications systems that operate in the license-free ISM band, ZigBee comes in on the lowball side as far as cost per node is concerned. As you read more and more about ZigBee and IEEE 802.15.4, think about what you've already read about Firefly. You'll experience déjà vu all over again.

The Muhammad Ali of Networks

I'm old. I still remember a young man from Louisville, Kentucky named Cassius Clay who stunned the world by knocking out a bigger and more powerful Sonny Liston. (Sonny was later implicated with the Mob. I wonder how much Mob money unexpectedly changed hands when Sonny's head hit the canvas.) Anyway, time passed and times changed. Cassius became Muhammad Ali, Sonny was purportedly employed by the Mob after retirement and the newly crowned king of boxing was to become by his poems an unexpected honorary ZigBee benefactor. Muhammad Ali once said (actually, he said it many times), "Float like a butterfly, sting like a bee." There we go again. Another reference to flying bugs. We all knew what Muhammad meant and you can apply his description of his fighting style directly to ZigBee. ZigBee is a lightweight wireless network protocol that packs a punch.

ZigBee is officially a wireless network protocol that is designed to be used with low-data-rate sensor and control networks. If a sensor doesn't need to report its condition constantly and allows the sensor-support electronics and radio to sleep most of the time, that sensor is what ZigBee was designed for. ZigBee can also eliminate the need to string wires all over the place, as it can easily reach data rates comparable to and above standard RS-232 and RS-485 wired protocols. Although an IEEE 802.15.4 network can easily obtain RS-232 speeds, you won't see many battery-powered applications of IEEE 802.15.4 networks replacing RS-232 communications links, especially if the traffic on the IEEE 802.15.4-based pseudo-RS-232 link is heavy. That's not what IEEE 802.15.4 networks or the ZigBee protocol is designed to do.

Let's get one item of confusion out of the way. ZigBee is not IEEE 802.15.4 and IEEE 802.15.4 is not ZigBee. ZigBee is a standards-based network protocol supported solely by the ZigBee Alliance that uses the transport services of the IEEE 802.15.4 network specification. The ZigBee Alliance is responsible for the ZigBee standard and the IEEE is boss when it comes to the IEEE 802.15.4 specification. That's no different from TCP/IP using the IEEE 802.11b wireless specification to transport IP datagrams and TCP segments between nodes equipped with 802.11b radios. The concept of a network protocol riding on a specification is called layering. Thus, ZigBee is layered on top of the IEEE 802.15.4 specification, just as TCP/IP can be layered over 802.11b or 802.3.

The IEEE 802.15.4 specification also uses internal layers, which are normally referred to as sublayers. The wireless 802.11b specification and the wired 802.3 specification also employ the concept of sublayers. The IEEE 802.15.4 specification calls out a pair of 802.15.4 sublayers, the PHY and the MAC.

The IEEE 802.15.4 standard actually defines not one but two PHYs, which span across three license-free frequency bands. One PHY spans the 868/915-MHz frequency band and the other PHY is dedicated to the 2.4-GHz frequency band. The 2.4-GHz frequency band supports a total of 16 channels numbered 11 to 26. Ten channels beginning with channel 1 and ending with channel 10 can be found in the 902-to-928-MHz frequency spectrum, which is commonly referred to as the 915-MHz band. The third license-free RF area, 868 to 870 MHz, only allows for channel 0. Data rates for each of the frequency domains also differ with the 2.4-GHz band allowing a maximum data rate of 250 kbps. The 902-to-928-MHz band data rate maximum is 40 kbps and the single-channel 868 to 870 MHz tops out at 20 kbps. Most everyone that walks the Earth can use the 2.4-GHz frequency spectrum. Only those that breathe air in North

America, Australia, New Zealand, Israel and Europe can operate their ZigBee radios in the bands below 1 GHz.

If we relate the IEEE 802.15.4 sublayers to the ZigBee protocol stack, the ZigBee PHY sublayer, which is actually the IEEE 802.15.4 PHY sublayer, is all about the radio and the generation of the radio link. A ZigBee stack's PHY responsibilities include receiver energy detection, link quality indication and clear channel assessment. The ZigBee stack's PHY is also primarily responsible for transmitting and receiving packets across the magnetic medium. The ability to sniff the air for other nodes is very important in the ZigBee and IEEE 802.15.4 world as this is what is done to determine if a new ZigBee or IEEE 802.15.4 network can be spawned.

The IEEE 802.15.4 MAC sublayer is in control of what is happening on the radio link. Again, relating the IEEE 802.15.4 MAC sublayer to the ZigBee stack, a ZigBee stack's MAC, which is actually the IEEE 802.15.4 MAC, is in control of the access to the radio channel and employs the services of CSMA-CA (Carrier Sense Multiple Access – Collision Avoidance) to avoid collisions on the radio link. Network association and disassociation are also duties that are handled by the ZigBee stack's MAC sublayer. And, if there is any security being applied to the radio link, the MAC is in charge of that too. Flow control, the acknowledgment and retransmission of data packets, frame validation and network synchronization falls on the IEEE 802.15.4 MAC sublayer in addition to everything else I've already mentioned. If all of that is not enough work for the MAC, it is also a key player in making sure the upper layers of the ZigBee stack get exactly what they require for proper application operation and data transfer.

If you had the chance to read my previous books on Ethernet networking, you know that we dealt with packet sizes that could extend beyond one thousand bytes. The maximum number of bytes we'll have to deal with in a ZigBee packet is only 127 and that includes the 16 bits of CRC (Cyclic Redundancy Check). The CRC has another name: *checksum*. The CRC bits make up a checksum value that is intended to help verify the integrity of the contents of the ZigBee or IEEE 802.15.4 packet. The IEEE 802.15.4 standard also includes an optional acknowledged data mode that allows any ZigBee or IEEE 802.15.4 network frame that has the ACK flag set to have its receipt acknowledged by the receiver. This is a good way to know if the frame has been received or not, but it doesn't guarantee that the frame will be processed by the receiver. The idea of the acknowledged frame is to continue to retry to deliver the frame after a defined timeout period. An error is declared if the frame is not acknowledged after a predefined number of delivery attempts. If the full packet addressing scheme is employed, the maximum amount of actual data a ZigBee packet can contain is 102 bytes.

ZigBee Devices

Before we can talk about how ZigBee networks can be configured, we must first understand who the ZigBee network players are and how they relate to IEEE 802.15.4. The IEEE 802.15.4 specification identifies two devices. They are the FFD and RFD. The 802.15.4 Full Function Device (FFD) is literally able to do it all. A typical FFD found in a ZigBee network will be powered from an inexhaustible power source, which is called out as an AC-fed mains supply, as it must always be active and listening on the network, among other things. An RFD (Reduced Function Device), on the other hand, is very limited in the tasks it can perform (recording temperature data, monitoring switches or controlling an external device) and more often than not depends on a battery for its power. Since the RFD's power source is easily exhaustible, an RFD is prone to sleep most of the time.

ZigBee takes the IEEE 802.15.4 concept of FFD and RFD and creates three ZigBee protocol devices. The ZigBee *Coordinator* is a one-of-a-kind FFD on the ZigBee network and actually forms the network. Once the Coordinator establishes a ZigBee network, it allocates network addresses for those that are allowed to join its network and maintains the binding table entries. A ZigBee Coordinator also routes messages between RFDs in a network.

A ZigBee *End Device* is the node that physically interfaces to a sensor or executes control functions. The ZigBee End Device can be either an FFD or RFD depending upon the End Device's intended application.

The third ZigBee device is optionally deployed. A ZigBee *Router* is an FFD that enables the extension of the physical range of a ZigBee network. The use of a ZigBee Router device allows more nodes to join the network as the radio range of the root ZigBee network is effectively increased. Since a ZigBee Router device is an FFD, it can also be used to perform End Device functions, such as monitoring sensors and executing control functions.

ZigBee Network Topologies

There are currently three common ZigBee network topologies. The *Star* network configuration is comprised of one ZigBee Coordinator and any number (within reason) of ZigBee End devices. The ZigBee End Devices are physically and electrically isolated from each other and depend on the ZigBee Coordinator to pass any type of information or message from ZigBee End Device to ZigBee End Device if that becomes necessary. As you already know, the ZigBee Coordinator is an FFD. The ZigBee End Devices can be FFDs or RFDs. ZigBee Star networks are termed single-hop networks as there is only one hop or path between a ZigBee End Device and the ZigBee Coordinator.

The ZigBee *Cluster Tree* topology employs the services of ZigBee End Devices that may join the ZigBee network via a ZigBee Coordinator or a ZigBee Router. In this topology the ZigBee Routers have two major functions. Since a ZigBee device can join the network via the router, more nodes may be able to participate in the Cluster Tree network. As I pointed out earlier, the ZigBee Router also has the ability to enhance the "reach out and touch someone" capability of the network. The inclusion of a ZigBee Router eliminates the need for an end device to be within radio range of the ZigBee network Coordinator. Just as in a ZigBee Star network, ZigBee End Devices cannot directly communicate with each other in a Cluster Tree network. ZigBee Routers speak directly to other ZigBee Routers and to ZigBee Coordinators. Messages from ZigBee End Devices that need to be routed to other ZigBee End Devices must be delivered by way of a ZigBee Router and/or a ZigBee Coordinator. So, a ZigBee Cluster Tree network is basically multiple ZigBee Star networks connected to the ZigBee Coordinator by ZigBee Routers. Since there are multiple paths that may be traversed to get a message to the ZigBee Coordinator, the ZigBee Cluster Tree topology is considered a multi-hop topology.

An extension to the ZigBee Cluster Tree topology is called the *Mesh* topology. ZigBee End Devices configured as RFDs are still forbidden to communicate directly with each other in a ZigBee Mesh network. However, ZigBee End Devices with FFD status can communicate with other FFDs without having to be routed through a ZigBee Router or ZigBee Coordinator. Just as you saw in the ZigBee Cluster Tree topology, ZigBee Routers and ZigBee Coordinators can make direct contact with each other if the devices are within radio range and ZigBee RFDs still must route their peer-to-peer messages through a ZigBee Coordinator or ZigBee Router.

Regardless of the ZigBee network topology, all ZigBee protocol networks allow each ZigBee node equal access to the network. Thus, a ZigBee Star network would be considered a single-hop multi-access network. Taking the multi-access concept to the other pair of ZigBee networks would declare Cluster Tree and Mesh ZigBee networks as multi-hop multi-access networks.

The multi-access method can possibly be used in two ways (nonBeacon and Beacon) by the ZigBee protocol. A nonBeacon-enabled network allows every node that is participating in the network to transmit at any time the channel is open. Nodes of a Beacon-enabled network can only transmit inside of a predetermined time slot. In a Beacon-enabled ZigBee network, the ZigBee Coordinator will periodically generate a superframe that is identified as a Beacon frame. Since the emission of the superframe implies the use of time slots, every node must synchronize with the superframe in the time domain. Each node participating in the Beacon method is assigned a specific time slot that it can use to transmit and receive data. Normally, a node in a Beacon-based network will synchronize with the ZigBee Coordinator's Beacons and wake up just before the Beacon is to be generated, hopefully do its thing inside the Beacon's active time period, and go back to sleep awaiting the next Beacon period. The superframe may also contain a time slot that is not dedicated to any specific node. In this wide-open time slot, any node that wants to transmit or receive must compete with other nodes with the same idea using the CSMA-CA methodology.

Patty Cake, Patty Cake

Baker's man, bake me a cake as fast as you can. Roll it and pat it and throw it in the PAN. That's my version of the classic. My point is that you have learned quite a bit about ZigBee and IEEE 802.15.4 networks, but there are still a few things I want to tell you before we move on.

When not in a baker's hands, a ZigBee/IEEE 802.15.4 network is also known as a PAN (Personal Area Network). You may also see a ZigBee/IEEE 802.15.4 network described as a WPAN (Wireless Personal Area Network) or LR-WPAN (Low Rate Wireless Personal Area Network). In your ZigBee travels, you may see the term PAN or WPAN Coordinator. No worries. That's the same thing as the ZigBee Coordinator we've already talked about. You may also see the terms ZC (ZigBee Coordinator), ZR (ZigBee Router) and ZED (ZigBee End Device). Just grin and know that you are on top of those ZigBee acronyms. You've also got a little IEEE 802.15.4 rubbed off on you, as you can now confidently discuss IEEE 802.15.4 network topologies.

Thus far, you have been introduced to the basics of ZigBee/IEEE 802.15.4 networking. We've discussed the physical layer (PHY) and the MAC (Medium Access Control) layer of the ZigBee stack model. We even ventured into a part of the NWK (Network) layer of a ZigBee stack when I presented a rundown of the different ZigBee network routing algorithms. The ZigBee stack's MAC and PHY layers are governed by the IEEE 802.15.4 specification. The ZigBee stack's NWK layer and all layers above it are under the control of the ZigBee Alliance specification. A typical ZigBee stack is organized in the fashion of Figure 1.1.

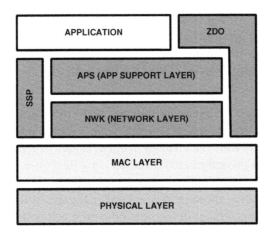

Figure 1.1: Typical ZigBee stack organization. This may look imposing right now, but as we move on up (as George Jefferson would say), you'll find it all to be quite logical.

The IEEE 802.15.4 specification is 679 pages in length. I don't plan on deciphering all of the IEEE 802.15.4 specification and I'm sure that you're not looking forward to me doing so. After all, this is a "hands-on" book. If there's anything more to ferret out that concerns IEEE 802.15.4, we'll dig it out and bring it to light as we discover the ways of ZigBee by way of practical experience or practical example.

Before we move on to bigger and better ZigBee and IEEE 802.15.4 things, I have a question for you. What do Floyd Council, Roger Barrett and Pinkney Anderson have in common? (The answer will be revealed as we move on...)

You Are Dangerous and You're Going to Hell

In the previous chapter you read and learned about certain elements of IEEE 802.15.4 and a little bit about the ZigBee protocol stack. You know just enough to be dangerous. If you're not already a ZigBee professional, at this point you only know enough about ZigBee and IEEE 802.15.4 to carry on a casual conversation at a cocktail party. However, since you put out your hard-earned money to buy this book, I feel that it is my duty to elevate your knowledge of the IEEE 802.15.4 standard and the ZigBee and IEEE 802.15.4 protocol definitions. In other words, I want you to be able to converse with the nerdiest of the nerds at the cocktail party.

The last thing I saw you do with your hands was play patty cake. As my Mama says, "Idle hands are the Devil's workshop." With the help of some advanced ZigBee and IEEE 802.15.4 tools, and some basic 802.15.4 hardware, I'm going to run you right through hell so fast that the Devil won't even know you were there. Before we start this trip, you've got to learn to walk on hot coals. Let's start at the bottom of the ZigBee stack fire and examine the inner workings of the ZigBee stack's PHY layer.

The IEEE 802.15.4 PHY

No matter where you go, there you are. Well, we are standing in the cellar of the ZigBee stack in a place called the PHY layer. Recall that I have warned you not to confuse what the IEEE folks do for ZigBee with what the ZigBee Alliance folks do with the upper layers of the Zig-Bee stack. The IEEE 802.15.4 PHY is governed by the IEEE 802.15.4 standards document. The PHY works for but does not directly report to the ZigBee specification layers, which reside above the IEEE 802.15.4 MAC sublayer.

You already know some basic things about the PHY. For instance, the PHY is responsible for the following tasks:

- Data transmission and reception

- CCA (Clear Channel Assessment) for
  CSMA-CA (Carrier Sense Multiple Access – Collision Avoidance)

- Activation and deactivation of the radio transceiver

- ED (Energy Detect) within the current channel

- Channel frequency selection

- LQI (Link Quality Indicator) for received packets

You also know that two PHYs, 868/915 MHz and 2.4 GHz, are defined by the IEEE 802.15.4 specification. To that end, the IEEE 802.15.4 standard is designed to conform with established regulations in Canada, Europe, Japan, Israel and the United States.

For all intents and purposes, the PHY's boss is the MAC, as everything the PHY does has to somehow flow through the MAC layer of the ZigBee stack. The PHY's job is to provide an interface between the MAC sublayer and the physical radio channel. There are actually three interlayer interfaces associated with the PHY, which are called SAPs (Service Access Points). The RF-SAP (Radio Frequency Service Access Point) is part physical and part logical as it is made up of the radio hardware and radio firmware. You will sometimes see IEEE 802.15.4 radio manufacturers call this the HAL (Hardware Abstraction Layer). A HAL is simply a piece of code that adapts the manufacturer's radio to the rest of the ZigBee/IEEE 802.15.4 node's circuitry. Like the HAL, the PD-SAP (PHY Data Service Access Point) and PLME-SAP (PHY Layer Management Entity – Service Access Point) access points are purely logical as they provide data and management access services between the PHY and MAC sublayers, respectively.

The PLME-SAP is the gateway to the PHY's PLME management entity. The PLME provides the layer management service interfaces, which allow the invocation of layer management functions. The PLME also has the responsibility of maintaining a database of managed PHY objects called the PHY PAN Information Base (PHY PIB). Rather than just throw out the PIB acronym, the bits and bytes that make up a PHY PIB can be seen in Code Snippet 2.1.

Code Snippet 2.1
****
```
typedef struct _PHY_PIB
{
    BYTE phyCurrentChannel;
    WORD phyChannelsSupported;
    union _phyTransmitPower
    {
        BYTE val;
        struct _phyTransmitPower_bits
        {
            unsigned int nominalTransmitPower :6;
            unsigned int tolerance :2;
        } bits;
    } phyTransmitPower;
    BYTE phyCCAMode;
} PHY_PIB;
```
****
Code Snippet 2.1: Here's a logical layout of a typical PHY PIB. It makes a lot more sense when you can visualize it, doesn't it?

The *phyCurrentChannel* PIB attribute can range from 0 to 26 and represents the RF channel to use for all following transmissions and receptions. The five most significant bits of *phyChannelsSupported* are reserved and will always be set to zero. That leaves the 27 least-significant bits of the *phyChannelsSupported* double word to indicate the availability status of each of the 27 channels. For instance, 0b0000 0000 0000 0000 0000 1000 0000 0000 tells us that channel 12 is available.

If you wear a pointy hat with stars and moons on it, and your RF (radio frequency) diploma/ certification was handed to you by an agent of the Devil in a blood red suit, you will be highly interested in the *nominalTransmitPower* and *tolerance* bitmap. The pair of most significant bits is representative of the tolerance on the transmit power, which ranges from ±1dB to ±3dB to ±6dB in three respective binary steps of 0b00, 0b01 and 0b10. The remaining six least-significant bits form a signed integer in twos-complement format that corresponds to the nominal transmit power of the device in decibels (dB) relative to 1 mW. Like I said, if you're an RF wizard, that's important. Otherwise, just keep all of that in memory so you can spit it out at the cocktail party. You know, baffle them with buffalo chips.

Three methods, or CCA Modes, apply to the *phyCCAMode* PIB attribute:

- CCA Mode 1 – Energy above threshold
- CCA Mode 2 – Carrier sense only
- CCA Mode 3 – Carrier sense with energy above threshold

CCA Mode 1 will return a busy medium status if any energy is detected above the ED threshold. Carrier sense only in the CCA Mode 2 short description says that the ED threshold is of no concern in the determination of a clear channel. Instead, CCA Mode 2 simply looks for a valid IEEE 802.15.4 signal and returns a busy medium status if a signal is detected. As you have already concluded, CCA Mode 3 is a combination of CCA Mode 1 and CCA Mode 2 as both a minimum ED threshold and the detection of a valid IEEE 802.15.4 signal will return a busy medium status. You're now officially trained on clear channel assessment.

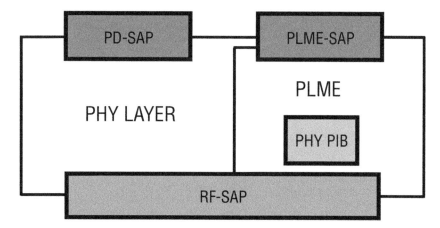

Figure 2.1: Reference model of PHY layer. Think of SAPs as simple transfer points between layers of the stack. The PHY PIB is actually a simple collection of data that can be accessed and used by the layer data-passing elements of the stack.

A CCA is performed when a PLME-CCA.request primitive is received by the PHY. The medium status of BUSY or IDLE is returned in a PLME-CCA.confirm primitive reply. What the heck is a primitive? Don't worry your pointy little head as we're about to talk about primitives. Meanwhile, a reference model of the PHY layer and all of its associated SAPs is posted for you in Figure 2.1.

The PHY Data Service

MAC protocol data units (MPDUs) flow between peer MAC sublayers by way of the PD-SAP courtesy of the PHY data service. The PD-SAP, as you'll discover with most other SAPs, supports a set of primitives. A primitive is simply a logical entity that conveys important information to an adjacent upper or lower layer by way of a SAP. Upper sublayers generally use the information passed to them by lower sublayers to use in their service to the next higher sublayer. In the case of the PHY data service, there are three PD-DATA primitives:

- PD-DATA.request
- PD-DATA.confirm
- PD-DATA.indication

Code Snippet 2.2 is a code visualization of the function provided by the PD-data primitives. Note that the PD-DATA primitives are simply data structures filled with parameters and do not carry any executable content within them.

Code Snippet 2.2

```
struct _PD_DATA_request
{
    BYTE    psduLength;
    BYTE    *psdu;
} PD_DATA_request;

struct _PD_DATA_confirm
{
    BYTE    status;
} PD_DATA_confirm;

struct _PD_DATA_indication
{
    BYTE    psduLength;
    BYTE    *psdu;
    BYTE    ppduLinkQuality;
} PD_DATA_indication;
```

Code Snippet 2.2: Get used to the term primitive *as it is used extensively in the ZigBee and IEEE 802.15.4 worlds.*

Normally, a request primitive is passed to request that a service be initiated. The confirm primitive will usually return the status of a service request such as SUCCESS or FAILURE.

An indication primitive signals that an event has occurred that most likely will need some attention. Confirm primitives also may provide a logical path to any data that was requested. Let's define the functionality of each of the PD-DATA, primitives beginning with PD-DATA. request.

The PD-DATA.request primitive requests the transfer of an MPDU (signified by *psdu* (PHY Service Data Unit) in Code Snippet 2.2) from the MAC sublayer to the PHY sublayer. If you examine the relationship between an MPDU and a PSDU, you'll see that the PSDU is simply an MPDU in the MAC's domain and vice versa with a MPDU being nothing more than a PSDU in the PHY's domain. Same data with a different name depending on which side of the MAC/PHY border you're standing on.

The PD-DATA.request parameters are to be used by the PHY and thus are communicated with PHY-related parameters. The number of bytes to be transmitted by the PHY is given by the *psduLength* parameter, while the actual packet to be transmitted is located in a memory space defined by the pointer *psdu*. Nothing fancy going here, as all we have thus far is probably no more than an array or data structure holding some data, a byte telling us how much data is there and a pointer to the beginning of the buffer (array or data structure) that is holding the data. Remember, we can only pack 127 bytes into an IEEE 802.15.4 packet. Thus, one byte will represent the length of an IEEE 802.15.4 packet with no problem.

When the PD-DATA.request primitive is pushed through the SAP and is received by the PHY, the PSDU will be transmitted. Before the PHY transmits the PSDU it received from the MAC layer, it will convert the incoming PSDU to a PPDU. If the transmitter is enabled, the PPDU will be transmitted. Take a look at Figure 2.2. A PPDU is simply a PSDU with the addition of the PHY header, which is comprised of a 4-byte preamble sequence, an 8-bit Start of Frame Delimiter and a frame length byte. The preamble sequence and the Start of Frame Delimiter as a unit are called the SHR (Synchronization Header). The single byte of frame length is also known as the PHR (PHY Header).

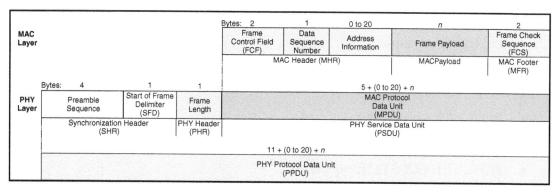

Figure 2.2: If you've had the time to read my other networking books, you're waiting for me to say the word encapsulation. *The Data Payload loses and gains header information as it traverses up and down the sublayer chain. The PSDU/MPDU relationship is obvious in this figure.*

Once the logic in the PHY and radio hardware have forced all of the bits out of the antenna, a PD-DATA.confirm primitive will be issued by the PHY with a status of SUCCESS. On the other hand, if a receive operation is in progress and the receiver is on or if the transmitter is off, the PD-DATA.confirm primitive will be issued with a RX_ON or TRX_OFF status, respectively.

If you take a close look at the PD-DATA.confirm primitive code in Code Snippet 2.2, you'll see that there is only a single parameter, *status*, in the PD-DATA.confirm primitive's data structure. As far as the PD-DATA.confirm primitive is concerned, there are only three states that the *status* parameter can take, which are SUCCESS, RX_ON and TRX_OFF. When the MAC layer receives the PD-DATA.confirm primitive, it gets the status of the PD-DATA.request primitive it previously issued to the PHY by examining the PD-DATA.confirm primitive's *status* byte.

The PD-DATA.indication primitive is issued by the PHY to transfer a received PSDU to the MAC sublayer. Note that the PHY packet terminology is used here, as the PHY will strip the PHY header and only transfer the PSDU to the MAC layer. Of course, as we have seen earlier, the MAC sees the PSDU and an MPDU in its domain. Upon arrival of the PD-DATA.indication primitive the MAC has access to the number of bytes contained within the incoming PSDU, the data payload within the PSDU and the link quality measured during the reception of the PPDU.

The PHY Management Service

If you understand how the PHY Data Service works, you'll have no problem with the PHY Management Service as it behaves at ground level in an almost identical manner. The PLME-SAP is the conduit in which the management commands flow between the MLME (MAC Sublayer Management Entity) and PLME (Physical Layer Management Entity).

Like the PHY Data Service, the PHY Management Service includes a SAP that is supported by primitives. The PLME-SAP supports the following primitives:

- PLME-CCA.request
- PLME-CCA.confirm
- PLME-ED.request
- PLME-ED.confirm
- PLME-GET.request
- PLME-GET.return
- PLME-SET-TRX-STATE.request
- PLME-SET-TRX-STATE.confirm
- PLME-SET.request
- PLME-SET.confirm

If you didn't doze off, you should be familiar with the acronyms and the mechanisms associated with request and confirm primitives. The PLME-CCA.request primitive requests that a CCA (Clear Channel Assessment) be performed by the PHY. This particular primitive has no parameters associated with it. The MLME simply issues the PLME-CCA.request primitive and an assessment of the channel is performed. The PLME will issue the PLME-CCA.confirm primitive in response to the MLME's PLME-CCA.request primitive. A status of BUSY or IDLE will be returned by the PLME-CCA.confirm primitive unless the transceiver is disabled or the transmitter is enabled, which will return a status of TRX_OFF and TX_ON, respectively.

To obtain an ED measurement, the MLME issues the PLME-ED.request primitive. Like the PLME-CCA.request primitive, the PLME-ED.request primitive has no parameters. The PHY will perform an ED measurement upon receipt of the PLME-ED.request primitive if the receiver is enabled. The PHY will issue a PLME-ED.confirm primitive with a status of SUCCESS if all of the ducks are in a row. Otherwise, TRX_OFF will be returned as a status if the transceiver is disabled and TX_ON will be returned if the radio is in the act of transmitting when the do-an-ED-operation command arrives. The energy level will be returned in the PLME-ED.confirm primitive as a number between 0x00 and 0xFF. A representation of a typical way to code the PLME-CCA and PLME-ED primitives is shown in Code Snippet 2.3.

Code Snippet 2.3

```
****************************************************************************
//      struct _PLME_CCA_request
//      {
//          No Inputs
//      } PLME_CCA_request;

    struct _PLME_CCA_confirm
    {
        BYTE    status;
    } PLME_CCA_confirm;

//      struct _PLME_ED_request
//      {
//          No Inputs
//      } PLME_ED_request;
    struct _PLME_ED_confirm
    {
        BYTE    status;
        BYTE    EnergyLevel;
    } PLME_ED_confirm;
****************************************************************************
```

Code Snippet 2.3: No rocket science here. Hopefully, you've grasped the idea of primitives.

Do you recall those PHY PIB attributes we discussed in Code Snippet 2.1? Well, here are the primitives that allow you to read and write them.

The MLME issues a PLME-GET.request primitive to the PLME specifying the PHY PIB attribute it wants to know more about. Upon receipt of the PLME-GET.request primitive, the PLME will attempt to retrieve the requested PHY PIB attribute from its database. The PLME will issue a PLME-GET.confirm primitive containing a status of UNSUPPORTED_ATTRIBUTE if the requested PIB attribute is not found within the PHY PIB database. Otherwise, the PLME will issue a PLME-GET.confirm primitive status of SUCCESS along with the PHY PIB identifier and the value of the PIB attribute if the PHY PIB value is retrieved.

As you would logically expect, the PLME-SET.request primitive kicks off an attempt by the PLME to write the requested PHY PIB attribute value to the associated PHY PIB attribute that should be residing in the PHY PIB database. Just like the PLME-GET.confirm primitive, the PLME-SET.confirm primitive will choke and return a status of UNSUPPORTED_ATTRIBUTE if the requested PHY PIB attribute is bogus and not found within the PHY PIB database. What one would want to see is the PHY PIB attribute identifier and SUCCESS in the status parameter of the PLME-SET.confirm primitive.

Code Snippet 2.4

```
*************************************************************************
    struct _PLME_GET_request
    {
        BYTE      PIBAttribute;
    } PLME_GET_request;

    struct _PLME_GET_confirm
    {
        BYTE      status;
        BYTE      PIBAttribute;
        void      *PIBAttributeValue;
    } PLME_GET_confirm;


    struct _PLME_SET_request
    {
        BYTE      PIBAttribute;
        void      *PIBAttributeValue;
    } PLME_SET_request;
    struct _PLME_SET_confirm
    {
        BYTE      status;
        BYTE      PIBAttribute;
    } PLME_SET_confirm;
*************************************************************************
```

Code Snippet 2.4: Odds are you'll not have to use all of these primitives. However, you'll sure look good and sound really smart at the cocktail party.

The PLME-SET_TRX-STATE primitives coded up in Code Snippet 2.5 are gnarley little buggers at first glance. The object of the PLME-SET_TRX-STATE primitives is to force the PHY to change the state of the transceiver. There are three possible transceiver scenarios:

- Transceiver Disabled (TRX_OFF)

- Transmitter Enabled (TX_ON)

- Receiver Enabled (RX_ON)

The MLME generates and issues the PLME-SET_TRX-STATE.request primitive and waits for the PLME to change the transceiver's state to that which was requested within the *state* parameter of the PLME-SET_TRX-STATE.request primitive. If everything goes as planned, our good old SUCCESS status is returned in the PLME-SET_TRX-STATE.confirm primitive. If the state that is requested is already in service, the PLME-SET_TRX-STATE.confirm primitive returns the current state.

Here's a rundown on what happens when somebody doesn't follow the PLME-SET_TRX-STATE.request primitive's script. If the PLME-SET_TRX-STATE.request primitive is issued with a state parameter of RX_ON or TRX_OFF and the PHY is transmitting a PPDU, BUSY_TX will be crammed into the PLME-SET_TRX-STATE.confirm primitive's status parameter slot. This isn't all bad since when the transmission is completed, unless something really stupid has gone down, the state of the transceiver will be changed.

If TX_ON or TRX_OFF states are desired and the PHY is receiving (RX_ON) beyond the SFD (Start of Frame Delimiter), BUSY_RX will be returned by the PLME-SET_TRX-STATE.confirm primitive and like the BUSY_TX situation, the ending is good if the monkey doesn't drop the wrench as the transceiver's change of state is deferred until the reception of the PPDU is complete. Any bits going out the antenna tube following the SFD constitutes a valid transmission in progress.

If I have to explain what happens when the FORCE_TX_OFF state is issued from within the PLME-SET_TRX-STATE.confirm primitive, donate this book to your local library. Code Snippet 2.5 contains the data structures that represent the PLMN-SET-TRX-STATE primitives.

Code Snippet 2.5

```
************************************************************************
    struct _PLME_SET_TRX_STATE_request
    {
        BYTE     state;
    } PLME_SET_TRX_STATE_request;


    struct _PLME_SET_TRX_STATE_confirm
    {
        BYTE     status;
    } PLME_SET_TRX_STATE_confirm;
************************************************************************
```

Code Snippet 2.5: Although the actions that are initiated by the primitives in the snippet are heady, the invocation of the primitives follows suit with everything you now know about primitives at this point.

If you're wondering how the status values within the primitives are conveyed, they aren't really words but a set of predefined enumerated values. I've laid them all out for you to see in Table 2.1.

Enumeration	Value	Description
BUSY	0 x 00	The CCA attempt has detected a busy channel.
BUSY_RX	0 x 01	The transceiver is asked to change its state while receiving.
BUSY_TX	0 x 02	The transceiver is asked to change its state while transmitting.
FORCE_TRX_OFF	0 x 03	The transceiver is to be switched off.
IDLE	0 x 04	The CCA attempt is to be switched off.
INVALID_PARAMETER	0 x 05	A SET/GET request was issued with a parameter in the primitive that is out of the valid range.
RX_ON	0 x 06	The transceiver is in or is to be configured into the receiver enabled state.
SUCCESS	0 x 07	A SET/GET, an ED operation, or a transceiver state change was successful.

Table 2.1: Status values and descriptions. Don't get too excited about keeping these in your head. These values are normally enumerated in the ZigBee or IEEE 802.15.4 network firmware.

Primitive Passing Technique

You've seen that primitives are really just data structures that relate logically to the service that they represent. If you're wondering how a primitive gets passed from layer to layer, here's how it's done within the Microchip ZigBee stack. Let's follow some incoming data through the PHY.

The Microchip ZigBee stack is always cycling through, looking for a primitive to act upon. To allow the ZigBee stack application to process them, every primitive used by the Microchip ZigBee stack is enumerated. That is, every primitive is assigned a unique number that corresponds to its name. I've listed only the primitives that relate to the PHY in Code Snippet 2.6. As we move into talking about the MAC and other parts of the ZigBee stack, I'll reveal the related primitives and their enumerated values accordingly in a need-to-know fashion. Note that the PD and PLME in the primitive names are directly related to the SAP (PD-SAP or PLME-SAP) that the primitive traverses.

Code Snippet 2.6
```
**********************************************************************
typedef enum _ZIGBEE_PRIMITIVE
{
    NO_PRIMITIVE = 0,
```

```
PD_DATA_request = 0x01,
PD_DATA_confirm = 0x02,
PD_DATA_indication = 0x03,
PLME_CCA_request = 0x04,
PLME_CCA_confirm = 0x05,
PLME_ED_request = 0x06,
PLME_ED_confirm = 0x07,
PLME_GET_request = 0x08,
PLME_GET_confirm = 0x09,
PLME_SET_TRX_STATE_request = 0x0A,
PLME_SET_TRX_STATE_confirm = 0x0B,
PLME_SET_request = 0x0C,
PLME_SET_confirm = 0x0D,
```
****

Code Snippet 2.6: These little boys and girls should be familiar as we've just examined them. Later on, you'll see that the MAC and other upper sublayers of the ZigBee stack also have a set of associated primitives, which are also enumerated.

If the NO_PRIMITIVE value is representing the next primitive to be acted upon, the Microchip ZigBee stack software is allowed to cycle through background and housekeeping tasks and search for the next primitive to process inside of the various task modules, which are associated with each layer of the Microchip ZigBee stack. Since at this juncture you're only familiar with the primitives that support the PHY, let's follow the PHYTasks path in Code Snippet 2.7.

Code Snippet 2.7

****

```
BOOL ZigBeeTasks( ZIGBEE_PRIMITIVE *command )
{
    ZigBeeStatus.nextZigBeeState = *command;

    do
    {
        CLRWDT();

        if(ZigBeeStatus.nextZigBeeState == NO_PRIMITIVE)
        {
            ZigBeeStatus.nextZigBeeState = PHYTasks(ZigBeeStatus.nextZigBeeState);
        }
        if(ZigBeeStatus.nextZigBeeState == NO_PRIMITIVE)
        {
            ZigBeeStatus.nextZigBeeState = MACTasks(ZigBeeStatus.nextZigBeeState);
        }
        if(ZigBeeStatus.nextZigBeeState == NO_PRIMITIVE)
        {
            ZigBeeStatus.nextZigBeeState = NWKTasks(ZigBeeStatus.nextZigBeeState);
        }
        if(ZigBeeStatus.nextZigBeeState == NO_PRIMITIVE)
```

```
{
    ZigBeeStatus.nextZigBeeState = APSTasks(ZigBeeStatus.nextZigBeeState);
}
if(ZigBeeStatus.nextZigBeeState == NO_PRIMITIVE)
{
    ZigBeeStatus.nextZigBeeState = ZDOTasks(ZigBeeStatus.nextZigBeeState);
}
```

Code Snippet 2.7: See how it works? NO_PRIMITIVE sends us into each of the task code modules, which may or may not generate a primitive that needs to be passed to another task module. If we find that we have to work on something that needs to be transferred up to the MAC inside of the PHYTasks module, we set up to pass the primitive inside the PHYTasks code using the ZigBeeStatus. nextZigBeeState. The next pass-through would then fall into the task module we specified in the ZigBeeStatus.nextZigBeeState and the processing would continue there.

Let's follow through as if a packet has just entered the PHY FIFO (First In First Out), which is used by the PHY to buffer incoming packets. The concept is more important than the detail behind the code. So, I'll do a bit of paraphrasing when it's appropriate.

In the case of the Microchip MRF24J40 IEEE 802.15.4 transceiver, the packet is received by the PHY and the MRF24J40 informs the PIC microcontroller of its arrival by issuing a hardware interrupt, which pulls the PSDU from the PHY FIFO and stores it into a buffer area that has been allocated in the PIC microcontroller's SRAM. The incoming PSDU must be processed quickly and we already know that the buffered PSDU needs to be passed to the MAC's MPDU area. Recall that the PD-DATA.indication primitive is issued by the PHY to transfer a received PSDU to the MAC sublayer. Access to the MAC sublayer is provided by the services of the PD-SAP.

Along the way, the PHYTask code has satisfied the need for a value for the *psdu* parameter within the PD-DATA.indication primitive by allocating a buffer area within the PIC microcontroller's SRAM and loading the *psdu* parameter with the beginning location of the PSDU buffer area. In the case of the Microchip ZigBee stack, the buffer area pointed to by the *psdu* parameter is loaded with the data contained in the incoming PSDU. In the meantime, the *psduLength* parameter value for the PD-DATA.indication primitive was also collected from the PPDU header and placed in the PD-DATA.indication primitive's data structure. The *ppduLinkQuality* value gets collected from the transceiver and is stored into the PD-DATA.indication primitive's data structure as well. Now that we have stuffed all of the data parameter values required by the PD-DATA.indication primitive onto their holes, we return the enumerated value of the PD-DATA.indication primitive (PD_DATA_indication = 0x03) to the caller, which will continue to cycle looking for the next primitive to act upon. That next primitive to be acted upon will be the PD-DATA.indication primitive, as the enumerated value of the PD-DATA.indication primitive will become the new value of the ZigBeeStatus.nextZigBeeState variable.

Some of the primitives in Code Snippet 2.8 aren't yet ripe as far as we're concerned because we haven't gone into any detail about the operation of the MAC. However, the point I want to make is evident in Code Snippet 2.8. You can call out the SAP used by each of the primitives (i.e., PD-SAP, PLME-SAP, MLME-SAP, MCPS-SAP) by simply noting the primitive's prefix (i.e., PD, PLME, MLME, MCPS). The primitive prefixes also tell all about the use of the primitive. For instance, MLME (MAC Sublayer Management Entity) primitives are used for MAC management tasks, while MCPS (MAC Common Part Sublayer) primitives are used in data transfer tasks.

You can also get a feel from Code Snippet 2.8 as to how the primitives flow. The MAC issues a PLME-CCA.request to the PHY and the PHY performs the action and returns the status to the MAC in a PLME-CCA.confirm primitive. The only SAP available to carry the aforementioned primitives between the PHY and MAC sublayers is the PLME-SAP and the prefixes of the primitives used are hollering PLME-SAP! PLME-SAP!

Code Snippet 2.8
****

```
switch(ZigBeeStatus.nextZigBeeState)
{
    // Check for the primitives that are handled by the PHY.
    case PD_DATA_request:
    case PLME_CCA_request:
    case PLME_ED_request:
    case PLME_SET_request:
    case PLME_GET_request:
    case PLME_SET_TRX_STATE_request:
        ZigBeeStatus.nextZigBeeState =
        PHYTasks(ZigBeeStatus.nextZigBeeState);
        break;

    // Check for the primitives that are handled by the MAC.
    case PD_DATA_indication:
    case PD_DATA_confirm:
    case PLME_ED_confirm:
    case PLME_GET_confirm:
    case PLME_CCA_confirm:
    case PLME_SET_TRX_STATE_confirm:
    case PLME_SET_confirm:
    case MCPS_DATA_request:
    case MCPS_PURGE_request:
    case MLME_ASSOCIATE_request:
    case MLME_ASSOCIATE_response:
    case MLME_DISASSOCIATE_request:
    case MLME_GET_request:
    case MLME_GTS_request:
    case MLME_ORPHAN_response:
    case MLME_RESET_request:
    case MLME_RX_ENABLE_request:
    case MLME_SCAN_request:
    case MLME_SET_request:
    case MLME_START_request:
```

```
case MLME_SYNC_request:
case MLME_POLL_request:
     ZigBeeStatus.nextZigBeeState =
     MACTasks(ZigBeeStatus.nextZigBeeState);
     break;
```

****

Code Snippet 2.8: Note the division of work done by the primitives. The layering as it pertains to primitives is very evident here. Use the prefix SAP clues to better understand who's zooming who over which SAP in MAC and PHY sublayer land.

The code flow will fall through the remainder of the code segment shown in Code Snippet 2.7 and pick up in the MACTasks module. This is the actual point of traversal of the PD-DATA. indication primitive and all of its associated parameters from the PHY sublayer to the MAC sublayer. Code within the MACTasks module will pick up on the PD_DATA_indication value (PD_DATA_indication = 0x03) and execute code to extract elements from the transferred MPDU and the passed primitive parameters that the MAC layer will use to service either the next higher layer or pass data back through the PHY for transmission. Tricky, but deathly simple, huh?

The Envelope, Please

If you answered, "They are all dead," you are correct, as they all have indeed passed away. You are also sorta correct if you pointed out that they all played the blues. I would have to argue with you about Syd on the blues point, as every Brit during that musical time period was a wannabe "blues man." However, the answer I'm looking for is found in the combination of the names of South Carolina's Pink Anderson and North Carolina's Floyd "Dipper Boy" Council, which were combined by Syd as a name for his new band, Pink Floyd.

Since we're on the dark side of life, here's one for you. What very famous deceased R&B star played with Neil Young in a band called the Mynah Birds?

Keep Running

Now you're even more dangerous than you were a chapter ago. That's just more cause for you to keep running hard. We've run right through the IEEE 802.15.4 PHY and the sign up ahead says "IEEE 802.15.4 MAC."

The MAC sublayer is in charge of any operation that involves the physical radio channel. In addition, the MAC is responsible for the tasks listed below:

- Providing a reliable link between two peer MAC entities

- Handling the CSMA-CA mechanisms for channel access

- PAN association and PAN disassociation

- Beacon synchronization

- Beacon generation

- Device security

- Overseeing the GTS mechanism

The MAC sublayer is very similar logically to the PHY sublayer. Naturally, the names have changed and what the mechanisms behind the names do and represent has also changed in the definition of the MAC sublayer. The duty of the MAC sublayer in the ZigBee domain is to provide the services I mentioned earlier and to serve as an interface between the ZigBee NWK sublayer and the PHY. The components that make up the IEEE 802.15.4 MAC are shown schematically in Figure 3.1.

Like the PHY sublayer, the MAC sublayer has a management entity, which is called the MLME (MAC Layer Management Entity). The MAC's MLME provides the same basic functionality as the PHY's PLME but in a MAC kind of way. That is, it provides service interfaces through which layer-management functions can be invoked. Remember the PHY PIB? Well, guess what. There's a MAC sublayer PIB, which resides inside of the MLME. A representation of the parameters found within the Microchip ZigBee stack MAC sublayer PIB appear in Code Snippet 3.1.

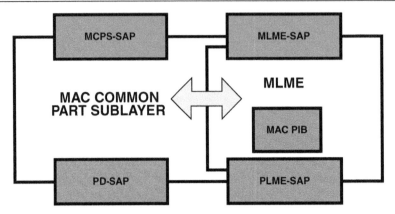

Figure 3.1: Components of IEEE 802.15.4 MAC. The SAPs are the portals to the upper and lower sublayers that the MAC is sandwiched between. Note also that a logical pipe has been laid between the MCPS and the MLME to allow the MLME to use the data services of the MCPS.

The MAC data service is accessed through the MCPS data SAP (MCPS-SAP). The MLME-SAP is used to provide access to the MAC management service. The data and management paths between the MAC sublayer and the PHY sublayer are completed by our old friends the PD-SAP and PLME-SAP, respectively. There is one additional interface used in the MAC sublayer that does not have an equal in the PHY sublayer. There is an implicit interface between the MLME and MCPS that allows the MLME access to the MAC data service.

We've already examined in detail one way primitives can be passed between layers via SAPs and in doing so we also revealed the logic behind a SAP. It will be a worthwhile exercise to examine each of the MAC PIB entries, as we will uncover other points of interest that will better our understanding of IEEE 802.15.4 operations.

The macAckWaitDuration value contained within Code Snippet 3.1 can range from 54 to 120 decimal. The value of macAckWaitDuration defines the maximum number of symbols to wait for an acknowledgment frame to be returned after a frame has been transmitted. The value of the macAckWaitDuration attribute is dependent upon the channel of operation. For channels 0 through 10, the macAckWaitDuration value should be equal to 120 decimal. The value of 54 decimal is common (and the IEEE 802.15.4 standard default value) for channels 11 through 26. As you can see in Code Snippet 3.1, the Microchip ZigBee wizards have set the macAckWaitDuration value to 57.

Code Snippet 3.1
****
```
typedef struct _MAC_PIB
{
//    BYTE macAckWaitDuration;   //made a constant
    unsigned int macAssociationPermit :1;
    unsigned int macAutoRequest :1;
    unsigned int macBattLifeExtPeriods :1;
```

```
      unsigned int macPromiscuousMode :1;
      unsigned int macRxOnWhenIdle :1;
      BYTE macBeaconPayload[MAC_PIB_macBeaconPayloadLength];
/* always 3 for non-Beacon, 5 for Beacon */
//    BYTE macBeaconPayloadLength;  /* made a constant */
//    BYTE macBeaconOrder;    //made a constant
      TICK macBeaconTxTime;
      BYTE macBSN;
      LONG_ADDR macCoordExtendedAddress;
      SHORT_ADDR macCoordShortAddress;
      BYTE macDSN;
      BYTE macMaxCSMABackoffs;
      BYTE macMinBE;
      PAN_ADDR macPANId;
      SHORT_ADDR macShortAddress;
//    BYTE macSuperframeOrder;  //made a constant
//    WORD macTransactionPersistenceTime; //made a constant
} MAC_PIB;
*****************************************************************
#define MAC_PIB_macAckWaitDuration (54+3)
#define MAC_PIB_macBeaconPayloadLength 3
#define MAC_PIB_macBeaconOrder 15
#define MAC_PIB_macSuperframeOrder 15
*********************************************************************
```

Code Snippet 3.1: If you actually read the PHY chapter, you already know what this is about. This sample MAC PIB structure was taken from the bowels of the Microchip ZigBee stack. Note that some of the MAC PIB parameters have been defined as constants by the Microchip ZigBee programmers.

If I were you, about now I'd be asking, "What the heck is a symbol?" For the answer, let's stop the truck, pull off the road and consult the IEEE 802.15.4 standard documentation.

Table 3.1 is a refresher view that lays out the three PHY frequency bands. Table 3.1 also specifies chip rates, bit rates and symbol rates for each frequency band.

PHY (MHz)	Frequency band (MHz)	Spreading parameters		Data parameters		
		Chip rate (kchip/s)	Modulation	Bit rate (kb/s)	Symbol rate (ksymbol/s)	Symbols
868/915	868–868.6	300	BPSK	20	20	Binary
	902–928	600	BPSK	40	40	Binary
2450	2400–2483.5	2000	O-QPSK	250	62.5	16-ary Orthogonal

Table 3.1: The three PHY frequency bands. Be aware that lots of the stuff you will read about ZigBee and IEEE 802.15.4 networks in other places assumes the operation of the network is within the 2.4-GHz band. That means the speeds and feeds you will be quoted may only be valid for 2.4 GHz networks and will not represent the speeds and feeds of the 868-MHz and 915-MHz bands.

Symbol	Chip sequence (C_0, C_1, C_2, ... , C_{31})
0	1 1 0 1 1 0 0 1 1 1 0 0 0 0 1 1 0 1 0 1 0 0 1 0 0 0 1 0 1 1 1 0
1	1 1 1 0 1 1 0 1 1 0 0 1 1 1 0 0 0 0 1 1 0 1 0 1 0 0 1 0 0 0 1 0
2	0 0 1 0 1 1 1 0 1 1 0 1 1 0 0 1 1 1 0 0 0 0 1 1 0 1 0 1 0 0 1 0
3	0 0 1 0 0 0 1 0 1 1 1 0 1 1 0 1 1 0 0 1 1 1 0 0 0 0 1 1 0 1 0 1
4	0 1 0 1 0 0 1 0 0 0 1 0 1 1 1 0 1 1 0 1 1 0 0 1 1 1 0 0 0 0 1 1
5	0 0 1 1 0 1 0 1 0 0 1 0 0 0 1 0 1 1 1 0 1 1 0 1 1 0 0 1 1 1 0 0
6	1 1 0 0 0 0 1 1 0 1 0 1 0 0 1 0 0 0 1 0 1 1 1 0 1 1 0 1 1 0 0 1
7	1 0 0 1 1 1 0 0 0 0 1 1 0 1 0 1 0 0 1 0 0 0 1 0 1 1 1 0 1 1 0 1
8	1 0 0 0 1 1 0 0 1 0 0 1 0 1 1 0 0 0 0 0 0 1 1 1 0 1 1 1 1 0 1 1
9	1 0 1 1 1 0 0 0 1 1 0 0 1 0 0 1 0 1 1 0 0 0 0 0 0 1 1 1 0 1 1 1
10	0 1 1 1 1 0 1 1 1 0 0 0 1 1 0 0 1 0 0 1 0 1 1 0 0 0 0 0 0 1 1 1
11	0 1 1 1 0 1 1 1 1 0 1 1 1 0 0 0 1 1 0 0 1 0 0 1 0 1 1 0 0 0 0 0
12	0 0 0 0 0 1 1 1 0 1 1 1 1 0 1 1 1 0 0 0 1 1 0 0 1 0 0 1 0 1 1 0
13	0 1 1 0 0 0 0 0 0 1 1 1 0 1 1 1 1 0 1 1 1 0 0 0 1 1 0 0 1 0 0 1
14	1 0 0 1 0 1 1 0 0 0 0 0 0 1 1 1 0 1 1 1 1 0 1 1 1 0 0 0 1 1 0 0
15	1 1 0 0 1 0 0 1 0 1 1 0 0 0 0 0 0 1 1 1 0 1 1 1 1 0 1 1 1 0 0 0

Table 3.2: PN sequences. Memorize the chip values you see here as you'll constantly be referencing them throughout the rest of this book. Better yet, just memorize the mathematical algorithms that derive the chip values. That's much easier. On a serious note, beware that the data symbol bits read from left to right instead of right to left in this table.

We'll begin by examining the symbol as it relates to the 2.4-GHz band of frequencies. We already know that the maximum data rate of an IEEE 802.15.4 PHY operating in the 2.4 GHz band is 250 kbps. We also know that a byte of information consists of eight bits, which can be divided into a pair of 4-bit nibbles. If the terms *byte* and *nibble* are familiar, you also know that each nibble can range from 0b0000 to 0b1111 and a byte includes the binary range of 0b00000000 through 0b11111111. The 2.4-GHz PHY's modulation scheme is based on a 16-ary quasi-orthogonal numbering system. That's another way of saying 16 nearly orthogonal pseudo-random (PN) noise sequences.

The PN sequences and their relative nibble values are presented in their entirety in Table 3.2. The four least-significant bits of each byte (or octet for the Mozart in you) represent a single data symbol. The same goes for the four most significant bits of that same byte. Then, each of the data symbols (upper nibble and lower nibble) is correlated to one of the 32-bit PN sequences you've memorized in Table 3.2. The bits are then shifted out of the antenna least-significant bit first after being half-sine pulse shaped and modulated using O-QPSK (Offset-Quadrature Phase-Shift Keying). If you are an alien from an advanced race, you are able to actually view the transmission in real time. For the technology-challenged Earthlings reading this book, I've provided a graphic of the transmission in Figure 3.2.

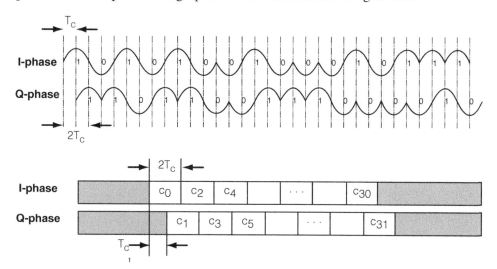

Figure 3.2: Graphic of data transmission. This is nice to know but you don't have to have in-depth knowledge of how the modulated bits look or hell from another planet to put an IEEE 802.15.4 network on the air.

Don't get too wrapped up in the complex wording as the science of O-QPSK is easily grasped. As you can see in Figure 3.2, each of those 32 bits in the pertinent chip value is modulated 90° out of phase, which is where the term quadrature comes into play. Quadrature actually means electrically 90° out of phase. Even count chips (c_0, c_2, c_4, etc.) are modulated on the in-phase (I) carrier while odd count chips (c_1, c_3, c_5, etc.) are modulated on the quadrature-phase (Q) carrier. For those of you that are not bona fide electrical engineers, the first 90° phase marker of chip c_0 in the I-phase is the beginning of chip c_1 in the Q-phase. The chips in the I-phase are leading the chips in the Q-phase by 90°. If your glass is half full, you would also be correct in saying that the chips in the Q-phase are lagging the chips in the I-phase by 90°. The 180° phase angle of chip c_0 occurs at the 90° peak of chip c_1. Got the idea?

To be able to obtain a symbol data rate of 62.5 Ksymbols/s at 2.4 GHz as specified in Table 3.1 requires that the chip rate be 32 times the symbol data rate. How did I get that relationship? Easy. One symbol is 32 chips. So, logic and Mr. Spock say that we must be able to transmit the 32 chips in the same time period we transmit the 4 bits, which is the same as one symbol. That equates to 32 times the symbol rate (32 * 62,500), which gives us a 2-Mchips/s chip rate to obtain a 62.5-Ksymbol/s symbol rate in the 2.4-GHz frequency band.

The chip-and-dip stuff is all fine and dandy but we still don't really know how long a symbol exists in the time domain. So, let's put our known values up against our unknown values in some simple mathematical equations to see if we can come up with a viable answer.

By way of the IEEE 802.15.4 standard document, we just discovered (we were actually told) that the symbol rate for a regulation 2.4-GHz IEEE 802.15.4 PHY is 62.5 Ksymbols/s. We also know that four bits (or 32 chips if you still have some dip left) make up a data symbol in the 2.4-GHz domain. The IEEE 802.15.4 standard document also tells us how the I/Q phase relationship is set and the width of each I-phase and Q-phase chip in the O-QPSK modulation scheme. The Q-phase chips are delayed by T_c with respect to the I-phase chips, where T_c is the inverse of the chip rate. Each chip is $2T_c$ wide. With what we know right now, we can mathematically deduce a symbol time unit (symbol period) and the value of T_c with some simple math:

> where:
> > ISM Band = 2.4 GHz
> > 250 kbps = maximum bit-per-second data rate
> > 62.5 Ksymbols/s = maximum symbol-per-second data rate
> > 4 = bits per symbol
> > symbol period = 1 / symbol frequency

> for a warm and fuzzy:
> > bits per symbol = 250 kbps / 62.5 Ksymbols/s = 4
> > symbol period = 1 / 62.5 Ksymbols/s = 16 μs
> > T_c = 1 / 2 Mchips/s = 500 ns

Things are a bit (pun intended) different with the lower-frequency PHYs. Table 3.3 shows us that each data bit is mapped into a 15-chip PN sequence instead of the 32-bit PN sequence we saw used in the 2.4-GHz band. BPSK (Binary Phase-Shift Keying) modulation with raised cosine pulse shaping is used in the 868-MHz and 915-MHz bands with chip rates of 300 kchips/s in the 868-MHz band and 600 kchips/s in the 915-MHz band.

Input bits	Chip values $(c_0 \; c_1 \; ... \; c_{14})$
0	1 1 1 1 0 1 0 1 1 0 0 1 0 0 0
1	0 0 0 0 1 0 1 0 0 1 1 0 1 1 1

Table 3.3: There's not as much to memorize here as there was for the 32-chip sequences. So, it's much easier to just put the 1's and 0's into your memory bank than to memorize the mathematical algorithms used to derive them. I guarantee that after a couple of cocktails, you'll be the hit of the party if you can cough up all 30 chip values correctly and find somebody that cares.

We already know what the maximum bit rates are for both of the lower-frequency PHYs by way of Table 3.1. So, we are capable of determining the bit rate using the chip rate and the number of bits per chip just as we did for the 2.4-GHz band. For instance, 15 chips represent one bit. Thus, the chip rate for the 868-MHz band is 15 times the bit rate. So, 300 kchips/s divided by 15 chips/bit is equal to 20 kbps. In this case, a symbol is equivalent to 1 bit. So, we can conclude that the symbol rate is 20 Ksymbols/s. We are correct, as the IEEE 802.15.4 standard document actually calls out the 20 Ksymbols/s figure in Table 3.1. It doesn't hurt to make sure we understand (and are speaking correctly at the cocktail party) by doing the simple math. Let's check our logic and, as Jethro Bodine would say, cipher on the rates for the 868-MHz and 915-MHz bands:

> where:
>> ISM Band = 868 MHz
>> 20 kbps = maximum bit-per-second data rate
>> 20 Ksymbols/s = maximum symbol-per-second data rate
>> 1 = bits per symbol
>> symbol period = 1 / symbol frequency

> for a warm and fuzzy:
>> bits per symbol = 20 kbps / 20 Ksymbols/s = 1
>> symbol period = 1 / 20 Ksymbols/s = 50 µs

> where:
>> ISM Band = 915 MHz
>> 40 Kbps = maximum bit-per-second data rate
>> 40 Ksymbols/s = maximum symbol-per-second data rate
>> 1 = bits per symbol
>> symbol period = 1 / symbol frequency

> for a warm and fuzzy:
>> bits per symbol = 40 kbps / 40 Ksymbols/s = 1
>> symbol period = 1 / 40 Ksymbols/s = 25 µs

As you can see, the bottom portion of Code Snippet 3.1 tells us that we aren't done with symbols. However, at least you know what a symbol is. If you decide to read the IEEE 802.15.4 specification, you'll find that your knowledge of symbols gained here will come in real handy. Let's get back in the truck and get on the road so we can continue our look at the MAC PIB attributes.

The next MAC PIB attribute listed in Code Snippet 3.1 is the macAssociationPermit bit. The macAssociationPermit is a Boolean attribute, which means it can take on a representative TRUE or FALSE condition. When the macAssociationPermit is TRUE, the coordinator is allowing other devices to associate.

We didn't get too far as we'll have to pull off the road again. The macAutoRequest attribute bit is also of the Boolean type. A device will automatically send a data request command if its address is listed in the Beacon frame and the macAutoRequest state is TRUE. I know…I know…What's a Beacon frame? OK, I'm pulling over.

The term *superframe* is used to describe the time interval that is bounded by a pair of consecutive Beacons. A Beacon is used to identify the PAN, convey network information, describe the structure of a superframe and synchronize devices attached to the network. Networks that operate with Beacons being emitted at predefined time intervals are called slotted networks, as designated time slots for devices to transfer data and commands are squeezed into the time between the emission of a pair of consecutive Beacons. Normally, in a slotted network a period of time called the Contention-Access Period (CAP) shares the superframe time with a Contention-Free Period (CFP) as shown in Figure 3.3.

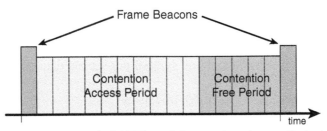

Figure 3.3: Contention-Free Period (CFP) and Contention-Access Period (CAP). The CAP always follows the Beacon and is the free-for-all area for devices on the network that wish to fight it out for air time. Only the privileged few get to party in the CFP.

The CAP is a time within the superframe in which any device can attempt to use the services of the PAN by way of a slotted CSMA-CA methodology. The CFP time within the superframe is made up of blocks of time called GTSs. A GTS (Guaranteed Time Slot) is dedicated to a particular application, which most likely maps to an associated device or set of devices. Time-critical applications will generally be prime candidates for GTS slots, as the opportunity to collect or exchange data is always guaranteed at the same time in each superframe cycle. The PAN Coordinator may allocate up to seven GTSs within a superframe. A GTS isn't limited to one superframe time slot and may extend over more than one superframe time slot period.

The coexistence of the CAP and CFP is possible within a superframe, as the GTS time domain is only allowed a portion of the time slots contained between the periodic Beacons of a superframe. The CAP time slots immediately follow the Beacon in slot 0 and precede the CFP area. All CAP activity must be completed before the CFP time period begins and all CFP activity must be completed within the allocated GTS timeframe and before the end of the CFP period, which is signaled by the emission of the next Beacon. It is possible to define a superframe that has no CFP functionality like the superframe shown in Figure 3.4.

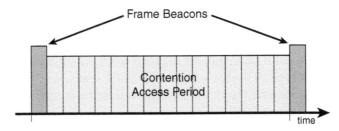

Figure 3.4: This is akin to one of those mountain-man-no-holds-barred wrestling matches. If you can get in the ring and do your thing, you win. There are no contenders coming into the ring riding in a limo.

It is also permitted to create a "dead zone" or inactive period within a superframe. This is accomplished by extending the Beacon-to-Beacon time to extend beyond the duration of the CAP and CFP periods. A logical view of the "dead zone" that can be carved out inside a superframe is shown in Figure 3.5.

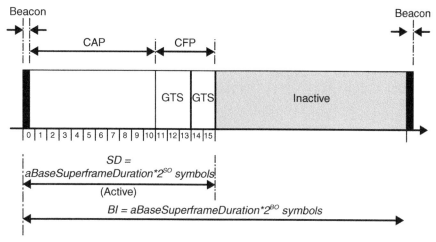

Figure 3.5: Adding a no-fly zone to a superframe allows everyone to sleep in if they are so inclined. The key to long-lived devices on IEEE 802.15.4 networks is their ability to conserve power whenever they can by sleeping.

The use of superframes is optional, which implies that an IEEE 802.15.4 network can operate without the presence of periodic Beacons. A network that does not require the services of continuous Beacons delineating superframes is called an unslotted network. The term *unslotted* is used to describe a network that does not allocate time slots, GTSs or generate periodic superframes using Beacons. Unslotted CSMA-CA techniques are employed in unslotted networks.

The best way to describe a Beacon frame is to examine a real one. A 10,000-foot view of a Beacon frame is drawn out for you in Figure 3.6.

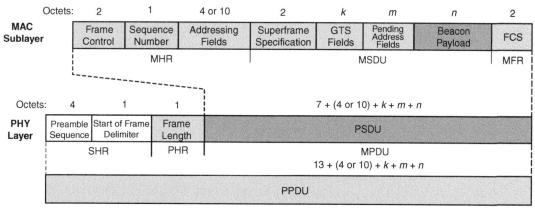

Figure 3.6: This view of a Beacon frame begins at the PHY level and ends up in the MAC sublayer. The acronyms should be logical and come to you easily by now.

The Daintree Networks SNA capture text shown in Capture Snippet 3.1 is a decoded 2.4-GHz band Beacon frame that was generated by the code within the Microchip ZigBee stack and put on the medium by a MRF24J40 transceiver.

Capture Snippet 3.1

```
***************************************************************************
Frame 3 (Length = 16 bytes)
        Time Stamp: 12:19:03.483
        Frame Length: 16 bytes
        Capture Length: 16 bytes
        Link Quality Indication: 136
IEEE 802.15.4
        Frame Control: 0x8000
                    .... .... .... .000  = Frame Type: Beacon (0x0000)
                    .... .... .... 0...  = Security Enabled: Disabled
                    .... .... ...0 ....  = Frame Pending: No more data
                    .... .... ..0. ....  = Acknowledgment Request: Acknowledgment
                                             not required
                    .... .... .0.. ....  = Intra PAN: Not within the PAN
                    .... ..00 0... ....  = Reserved
                    .... 00.. .... ....  = Destination Addressing Mode: PAN
                                             identifier and address field are not
                                             present (0x0000)
                    ..00 .... .... ....  = Reserved
                    10.. .... .... ....  = Source Addressing Mode: Address field
                                             contains a 16-bit short address
                                             (0x0002)
        Sequence Number: 152
        Source PAN Identifier: 0x0744
        Source Address: 0x0000
        MAC Payload
                Superframe Specification: 0xcfff
                    .... .... .... 1111  = Beacon Order (0x000f)
                    .... .... 1111 ....  = Superframe Order (0x000f)
                    .... 1111 .... ....  = Final CAP Slot (0x000f)
                    ...0 .... .... ....  = Battery Life Extension: Disabled
                    ..0. .... .... ....  = Reserved
                    .1.. .... .... ....  = PAN Coordinator: Transmitter is
                                             a PAN Coordinator
                    1... .... .... ....  = Association Permit: Coordinator
                                             accepting Association Requests
                GTS Specification: 0x00
                    .... .000  = GTS Descriptor Count (0x00)
                    .000 0...  = Reserved
                    0... ....  = GTS Permit: Coordinator not accepting GTS
                                    Requests
                Pending Address Specification: 0x00
                    .... .000  = Number of short Addresses pending: 0
                    .... 0...  = Reserved
                    .000 ....  = Number of extended Addresses pending: 0
                    0... ....  = Reserved
```

```
        Beacon Payload
              Protocol ID: ZigBee NWK (0x00)
     Frame Check Sequence: Correct
NWK Layer Information: 0x8411
     .... .... .... 0001  = Stack Profile (0x1)
     .... .... 0001 ....  = nwkcProtocolVersion (0x1)
     .... ..00 .... ....  = Reserved (0x0)
     .... .1.. .... ....  = Router Capacity: True
     .000 0... .... ....  = Device Depth (0x0)
     1... .... .... ....  = End Device Capacity: True

0000:   00 80 98 44 07 00 00 ff cf 00 00 00 11 84 .. ..    ...D....O.......
```
****

Capture Snippet 3.1: Every little bit that the MRF24J40 2.4-GHz transceiver slung out there was caught and analyzed by the Daintree Networks SNA application.

Let's tear the capture apart piece by piece beginning with the information gleaned from the PHR.

According to Capture Snippet 3.1.1, the Beacon frame consists of 16 bytes, including a pair of FCS characters contained within the MFR (MAC Footer). Note that the PHR containing the frame length value and the PSDU are the only survivors following the PHY's transfer of the PSDU to the MAC sublayer.

Capture Snippet 3.1.1
****
```
Frame 3 (Length = 16 bytes)
     Time Stamp: 12:19:03.483
     Frame Length: 16 bytes
     Capture Length: 16 bytes
     Link Quality Indication: 136
```
****
Capture Snippet 3.1.1

The contents of the MHR (MAC Header) lie within the bounds of Capture Snippet 3.1.2. The MHR is populated with two bytes of Frame Control information, a Beacon Sequence Number and, in this case, four bytes of Source Address Information. The Frame Control word identifies the frame as a Beacon frame that is using the 16-bit short addressing mode in lieu of the official 64-bit IEEE addressing mode. You can also see that the PAN is identified with a unique 16-bit address. You can find the PAN Identifier value in the MAC PIB disguised as macPANid. The Source Address is stored in the MAC PIB and resides inside the macCoord-ShortAddress variable.

Although we already know where a Beacon should come from, if we didn't, we could make the assumption that this Beacon was issued by a Coordinator, as the Source Address is 0x0000. The good news is that we don't have to assume (that turns one into a donkey). The device that transmitted the Beacon is identified later in the Beacon bit stream.

The unassuming Sequence Number value is also taken from the MAC PIB. The Sequence Number value is kept in the MAC PIB's macBSN variable.

Capture Snippet 3.1.2
****
```
IEEE 802.15.4
        Frame Control: 0x8000
                        .... .... .... .000 = Frame Type: Beacon (0x0000)
                        .... .... .... 0... = Security Enabled: Disabled
                        .... .... ...0 .... = Frame Pending: No more data
                        .... .... ..0. .... = Acknowledgment Request: Acknowledgment
                                                  not required
                        .... .... .0.. .... = Intra PAN: Not within the PAN
                        .... ..00 0... .... = Reserved
                        .... 00.. .... .... = Destination Addressing Mode: PAN
                                                  identifier and address field are not
                                                  present (0x0000)
                        ..00 .... .... .... = Reserved
                        10.. .... .... .... = Source Addressing Mode: Address field
                                                  contains a 16-bit short address
                                                  (0x0002)
        Sequence Number: 152
        Source PAN Identifier: 0x0744
        Source Address: 0x0000


0000:    00 80 98 44 07 00 00 ff cf 00 00 00 11 84 .. ..      ...D....O.......
```
****
Capture Snippet 3.1.2

The first two bytes of the MAC Payload in Capture Snippet 3.1.3 give us some information about what the superframe will look like and who transmitted this Beacon. We've already discussed the macAssociationPermit attribute and you can see now where it is used.

Capture Snippet 3.1.3
****
```
        MAC Payload
            Superframe Specification: 0xcfff
                        .... .... .... 1111 = Beacon Order (0x000f)
                        .... .... 1111 .... = Superframe Order (0x000f)
                        .... 1111 .... .... = Final CAP Slot (0x000f)
                        ...0 .... .... .... = Battery Life Extension: Disabled
                        ..0. .... .... .... = Reserved
                        .1.. .... .... .... = PAN Coordinator: Transmitter is
                                                  a PAN Coordinator
                        1... .... .... .... = Association Permit: Coordinator
                                                  accepting Association Requests
            GTS Specification: 0x00
                        .... .000 = GTS Descriptor Count (0x00)
                        .000 0... = Reserved
                        0... .... = GTS Permit: Coordinator not accepting GTS
                                         Requests
            Pending Address Specification: 0x00
                        .... .000 = Number of short Addresses pending: 0
                        .... 0... = Reserved
```

```
                          .000 ....  = Number of extended Addresses pending: 0
                          0... ....  = Reserved

0000:    00 80 98 44 07 00 00 ff cf 00 00 00 11 84 .. ..      ...D....o.......
```

#define MAC_PIB_macBattLifeExt FALSE
****
Capture Snippet 3.1.3

The Battery Life Extension, which is disabled in Capture Snippet 3.1.3, tells us if the coordinator receiver operation time is reduced during the CAP to conserve battery life. You don't see the macBattLifeExt variable in our MAC PIB lineup as it is defined as FALSE in the Microchip ZigBee stack's zigbee.def module. What you do see in the MAC PIB data structure is macBattLifeExtPeriods, which can take on values of 6 or 8. The macBattLifeExtPeriods variable indicates the number of backoff periods that the receiver is enabled following a Beacon in battery-life extension mode. If battery-life extension mode is enabled, the value of macBattLifeExtPeriods is set to 8 for channels 0 through 10 and set to a value of 6 for channels 11 through 26. There's your battery-life trivia question for the cocktail party.

The Beacon Order subfield of the Superframe Specification field is used to specify how often a Beacon is transmitted. When computing the actual elapsed times, Beacon Order is specified as BO along with a Beacon Interval, which is represented by BI. BI is equal to aBaseSuperframeDuration * 2^{BO} symbols, where BO is greater than or equal to 0 (zero) and less than or equal to 14.

One of many MAC sublayer constants defined within the IEEE 802.15.4 standard, aBaseSuperframeDuration represents the number of symbols forming a superframe when the superframe order is equal to 0 (zero). We'll get to a definition of superframe order in a moment. First, let's finish up our aBaseSuperframeDuration discussion.

The MAC sublayer constant aBaseSuperframeDuration is derived as follows:

aBaseSuperframeDuration = aBaseSlotDuration * aNumSuperframeSlots

Where:
aBaseSlotDuration = 60
aNumSuperframeSlots = 16

The number of symbols that make up a superframe slot when the superframe order is equal to zero is the official IEEE 802.15.4 standard definition of the MAC sublayer constant aBaseSlotDuration. Sixteen slots make up the number of slots in any superframe and that number of slots is given as aNumSuperframeSlots in the IEEE 802.15.4 standard documentation. We have already determined a symbol is 16 µs in length for the 2.4-GHz band, 25 µs long in the 915-MHz band and 50 µs long in the 868-MHz band. So, some simple substitution reveals the relative value of aBaseSuperframeDuration:

Where: ISM Frequency Band = 2.4 GHz
aBaseSuperframeDuration = 60 * 16 * 16 µs = 15.36 ms

Where: ISM Frequency Band = 915 MHz
aBaseSuperframeDuration = 60 * 16 * 25 μs = 24.00 ms

Where: ISM Frequency Band = 868 MHz
aBaseSuperframeDuration = 60 * 16 * 50 μs = 48.00 ms

The science and elegance of mathematics sometimes only reflect what should exist in a per-fect world. If I could point you to a perfect real-world mathematical model of every element of IEEE 802.15.4 and ZigBee, there would be no need for this book. You like "touchy-feely" just as much as I do and that's why this book exists (and why you're reading it). So, I put the Daintree Networks SNA and a couple of ZMD 900-MHz radios tuned to channel 1 (906 MHz) to work on this one using a cross-section of BO (Beacon Order) and SO (Superframe Order) value combinations. Here's what I came up with:

Time	Time Delta	MAC Src	Protocol	Packet Type
12:25:12.642	+00:00:00.024	0x5353	IEEE 802.15.4	Beacon: BO: 0, SO: 0
12:25:12.666	+00:00:00.024	0x5353	IEEE 802.15.4	Beacon: BO: 0, SO: 0
12:25:12.690	+00:00:00.024	0x5353	IEEE 802.15.4	Beacon: BO: 0, SO: 0
12:25:12.715	+00:00:00.024	0x5353	IEEE 802.15.4	Beacon: BO: 0, SO: 0
12:14:10.000		0x5353	IEEE 802.15.4	Beacon: BO: 5, SO: 4
12:14:10.777	+00:00:00.777	0x5353	IEEE 802.15.4	Beacon: BO: 5, SO: 4
12:14:11.553	+00:00:00.776	0x5353	IEEE 802.15.4	Beacon: BO: 5, SO: 4
12:14:12.329	+00:00:00.776	0x5353	IEEE 802.15.4	Beacon: BO: 5, SO: 4
13:33:49.052		0x5353	IEEE 802.15.4	Beacon: BO: 5, SO: 0
13:33:49.828	+00:00:00.777	0x5353	IEEE 802.15.4	Beacon: BO: 5, SO: 0
12:26:31.000		0x5353	IEEE 802.15.4	Beacon: BO: 14, SO: 13
12:33:08.458	+00:06:37.458	0x5353	IEEE 802.15.4	Beacon: BO: 14, SO: 13
13:22:32.000		0x5353	IEEE 802.15.4	Beacon: BO: 14, SO: 0
13:29:09.464	+00:06:37.464	0x5353	IEEE 802.15.4	Beacon: BO: 14, SO: 0

Here are the ideal mathematical Beacon Intervals for the 915-MHz frequency band:

Where: $BI = aBaseSuperframeDuration * 2^{BO}$ symbols

- When BO = 0, BI = 1 * 24.00 ms = 24.00 ms

- When BO = 5, BI = 32 * 24.00 ms = 768.00 ms

- When BO = 14, BI = 16384 * 24.00 ms = 393.22 S = 6.55 minutes

This book is not the result of or part of any government contract. However, the numbers gathered by the Daintree Networks SNA versus the ideal mathematical values are good enough for government work. What I really want you to come away with here is that the BO value determines the time interval between periodic Beacons. Note the Beacon interval time deltas for BO = 5 and SO = 0 are equal to the Beacon interval time deltas when BO = 5 and SO = 4.

There was a reason for calculating the aBaseSuperframeDuration values in the way we did earlier. The 24.00-ms value for the 915-MHz band is an ideal value that is based on symbols when SO = 0 (zero). With that understood, the value of SO is used in the determination of the superframe duration, which is dubbed SD. SD is the length of the active time of a superframe inside of the BI. The value of SD is calculated as follows:

SD = aBaseSuperframeDuration * 2^{SO}

Thus, we can mathematically conclude that for the 915-MHz frequency band:

- When SO = 0, SD = 1 * 24.00 ms = 24.00 ms

- When SO = 4, SD = 16 * 24.00 ms = 387.00 ms

- When SO = 13, SD = 8192 * 24.00 ms = 196.61 S = 3.28 minutes

It makes sense that when the superframe order is equal to the Beacon order, the duration of the superframe (SD) is equal to the Beacon interval (BI). Hopefully, you can also use the SD and BI numbers to get a feel for the relationship between the "dead zone" time versus the active superframe time within a Beacon interval.

To further prove out our symbolic math and the accuracy of the radio design versus the IEEE 802.15.4 standard, I tuned my network, the Daintree Networks SNA and the ZMD ZMD44102 capture device to the European 868-MHz band and captured the IEEE 802.15.4 bits as they flowed from antenna to antenna. Here are the results of the 868-MHz capture with BO and SO both equal to zero:

Time	Time Delta	MAC Src	Protocol	Packet Type
17:53:52.000		0x5353	IEEE 802.15.4	Beacon: BO: 0, SO: 0
17:53:52.049	+00:00:00.049	0x5353	IEEE 802.15.4	Beacon: BO: 0, SO: 0
17:53:52.097	+00:00:00.049	0x5353	IEEE 802.15.4	Beacon: BO: 0, SO: 0

Again, this is close enough for government work. Before we make the capture and take the measurement, let's calculate what the Beacon interval should be when BO = 5 at 868 MHz:

Where:

BI = aBaseSuperframeDuration * 2^{BO} symbols
BI = 48.00 ms * 32 = 1.54 s

Here's what the Daintree Networks SNA capture had to say:

Time	Time Delta	MAC Src	Protocol	Packet Type
18:10:16.000		0x5353	IEEE 802.15.4	Beacon: BO: 5, SO: 0
18:10:17.553	+00:00:01.553	0x5353	IEEE 802.15.4	Beacon: BO: 5, SO: 0
18:10:19.106	+00:00:01.553	0x5353	IEEE 802.15.4	Beacon: BO: 5, SO: 0

I couldn't crank the ZMD ZMD44102 868/900-MHz radios I used for this segment up to the 2.4-GHz band. So, I reached into my box full of Atmel AT86RF230 2.4-GHz transceiver boards and pulled one out. I cranked in a superframe order and Beacon order of zero. You know what that means:

Where: BI = aBaseSuperframeDuration * 2^{BO} symbols

- When BO = 0, BI = 1 * 15.36 ms = 15.36 ms

Here's what the Daintree Networks SNA capture had to say:

Time	Time Delta	MAC Src	Protocol	Packet Type
13:55:52.000		0xbabe	IEEE 802.15.4	Beacon: BO: 0, SO: 0
13:55:52.015	+00:00:00.015	0xbabe	IEEE 802.15.4	Beacon: BO: 0, SO: 0
13:55:52.031	+00:00:00.015	0xbabe	IEEE 802.15.4	Beacon: BO: 0, SO: 0

Then, just for grins, I dialed in a superframe and Beacon order of 4. If everything I've told you thus far is true (and you know it is), then the math should prove out against the Daintree Networks SNA capture data here as well:

Where: BI = aBaseSuperframeDuration * 2^{BO} symbols

- When BO = 4, BI = 16 * 15.36 ms = 245.76 ms

The envelope please:

Time	Time Delta	MAC Src	Protocol	Packet Type
14:52:27.867	+00:00:00.246	0xbaad	IEEE 802.15.4	Beacon: BO: 4, SO: 4
14:52:28.112	+00:00:00.246	0xbaad	IEEE 802.15.4	Beacon: BO: 4, SO: 4

Close enough for government work.

Here are the rules of engagement we've followed as they pertain to SO and BO so far:

- SO must always be less than or equal to BO

- Active SO and BO values range between 0 and 14

- SO and BO are don't cares if their values equal 15

You're going to love this. As far as the Beacon capture we started with is concerned, all of that superframe math we just performed was simply an exercise, as both the Beacon Order and Superframe Order fields contain a decimal 15. What this means is that the Coordinator will not transmit a Beacon frame unless it is requested to do so. For instance, a Beacon request command will trigger a Beacon frame as a response and that's exactly what happened in this case. The decimal 15 in the Superframe Order field deactivates the superframe that would normally immediately follow the Beacon. Note that this version of the Microchip ZigBee stack does not support slotted networks as both the Beacon Order and Superframe Order fields are permanently defined at decimal 15, as shown in Code Snippet 3.2.

Code Snippet 3.2
****

```
#define MAC_PIB_macBeaconOrder 15
#define MAC_PIB_macSuperframeOrder 15
```
****

Code Snippet 3.2: This set of 15's is a surefire way to identify an IEEE 802.15.4 unslotted network. No automatic Beacons, no GTSs and no superframes will be anywhere to be found.

Since the superframe has been eliminated from the network we captured this Beacon from, the GTS area does not exist and the gaggle of zeroes in the GTS Specification fields in Capture Snippet 3.1.4 drives that home.

Capture Snippet 3.1.4
****

```
        MAC Payload
            Superframe Specification: 0xcfff
                    .... .... .... 1111  = Beacon Order (0x000f)
                    .... .... 1111 ....  = Superframe Order (0x000f)
                    .... 1111 .... ....  = Final CAP Slot (0x000f)
                    ...0 .... .... ....  = Battery Life Extension: Disabled
                    ..0. .... .... ....  = Reserved
                    .1.. .... .... ....  = PAN Coordinator: Transmitter is
                                             a PAN Coordinator
                    1... .... .... ....  = Association Permit: Coordinator
                                             accepting Association Requests
            GTS Specification: 0x00
                    .... .000  = GTS Descriptor Count (0x00)
                    .000 0...  = Reserved
```

```
                        0... ....  = GTS Permit: Coordinator not accepting GTS
                                        Requests
              Pending Address Specification: 0x00
                        .... .000  = Number of short Addresses pending: 0
                        .... 0...  = Reserved
                        .000 ....  = Number of extended Addresses pending: 0
                        0... ....  = Reserved
```

```
0000:   00 80 98 44 07 00 00 ff cf 00 00 00 11 84 .. ..      ...D....O.......
```
****

Capture Snippet 3.1.4

The last three bytes of our Beacon are optional and the presence of the MAC_PIB_macBeaconPayloadLength definition in Capture Snippet 3.1.5 assures their presence.

Capture Snippet 3.1.5
****

```
         Beacon Payload
                 Protocol ID: ZigBee NWK (0x00)
     Frame Check Sequence: Correct
NWK Layer Information: 0x8411
         .... .... .... 0001  = Stack Profile (0x1)
         .... .... 0001 ....  = nwkcProtocolVersion (0x1)
         .... ..00 .... ....  = Reserved (0x0)
         .... .1.. .... ....  = Router Capacity: True
         .000 0... .... ....  = Device Depth (0x0)
         1... .... .... ....  = End Device Capacity: True
```

```
0000:   00 80 98 44 07 00 00 ff cf 00 00 00 11 84 .. ..      ...D....O.......
```

```
#define MAC_PIB_macBeaconPayloadLength 3
```
****

Capture Snippet 3.1.5

The data contained within the three optional bytes comes from the three elements of the macBeaconPayload[] array. It's rather easy to see in Code Snippet 3.3 how the three octets work their way into their spots in the Beacon Payload fields. The contents of the Beacon Payload field are intended for use by sublayers above the MAC sublayer.

Code Snippet 3.3
****

```
#define ZIGBEE_PROTOCOL_ID       0x00
#define nwkcProtocolVersion      0x01
#define MY_STACK_PROFILE_ID      0x01

void SetBeaconPayload( void )
{
    macPIB.macBeaconPayload[0] = ZIGBEE_PROTOCOL_ID;
    macPIB.macBeaconPayload[1] = (nwkcProtocolVersion  << 4) |
      MY_STACK_PROFILE_ID;
```

```
    macPIB.macBeaconPayload[2] = currentNeighborTableInfo.depth << 3;

if (macPIB.macAssociationPermit &&
    (currentNeighborTableInfo.numChildren < NIB_nwkMaxChildren) &&
    (currentNeighborTableInfo.neighborTableSize < MAX_NEIGHBORS))
{
    if (!nwkStatus.flags.bits.bAllEndDeviceAddressesUsed)
    {
        macPIB.macBeaconPayload[2] |= 0x80; // End Devices can join
    }
    if ((currentNeighborTableInfo.depth < (NIB_nwkMaxDepth-1)) &&
        (currentNeighborTableInfo.numChildRouters < NIB_nwkMaxRouters) &&
        !nwkStatus.flags.bits.bAllRouterAddressesUsed)
    {
        macPIB.macBeaconPayload[2] |= 0x04; // Routers can join
    }
}
}
```

****

Code Snippet 3.3: If you take your time and read through the long "if" arguments, you'll see that this code segment is simply telling the upper layers that end device and router addresses are available for use.

I searched through the Microchip ZigBee stack and didn't find any other occurrences of the MAC PIB elements macPromiscuousMode or macRxOnWhenIdle. Their names give away their purpose. The MAC is in a receive all mode when macPromiscuousMode is TRUE. When macRxOnWhenIdle is TRUE, the MAC sublayer enables the receiver during idle periods.

I didn't find that the MAC PIB element macBeaconTxTime had been used anywhere in the Microchip stack either. That would make sense as this parameter is used in slotted networks. Recall that the Microchip ZigBee stack at this time does not support slotted networks. The macBeaconTxTime is a 20-bit symbol time indicating when the last Beacon frame was transmitted. As you would imagine, the time is taken at the same point within every transmitted Beacon frame.

The macDSN is much like the macBSN in that it gets incremented and added to every transmitted data or MAC command frame. The macBSN is the macDSN's counterpart in a Beacon frame.

Code Snippet 3.3 is an excerpt from the Microchip ZigBee stack source code that relates how the macMinBE and macMaxCSMABackoffs MAC PIB variables are used. Let's suffice to say that macMinBE is the minimum value of the backoff exponent in the CSMA-CA algorithm and leave it at that for now. The default value of macMinBE is 3 according to the IEEE 802.15.4 standard document and that is reflected in Code Snippet 3.4. The maximum number of back-offs the CSMA-CA algorithm will tolerate before signaling a channel access failure is set by the macMaxCSMABackoffs value. Once again, the default value declared for the macMaxCSMABackoffs in the IEEE 802.15.4 standard document has been used here.

Code Snippet 3.4

****

```
            if(params.MLME_RESET_request.SetDefaultPIB==TRUE)
            {
                macPIB.macPANId.Val = 0xFFFF;
                macPIB.macShortAddress.Val = 0xFFFF;
                phyPIB.phyCurrentChannel = 11;
                macPIB.macAssociationPermit = FALSE;
                macPIB.macMaxCSMABackoffs = 4;
                macPIB.macMinBE = 3;
            }
```

****

Code Snippet 3.4: I don't know about you, but the smoke just seems to clear when I match up variables and their functions.

The macTransactionPersistenceTime is defined in superframe periods as the maximum time a transaction can be stored by a coordinator and indicted in its Beacon. The IEEE 802.15.4 default for macTransactionPersistenceTime is 0x01F4 superframe periods.

Tired Yet??

You better not poop out on me. We're just getting started. I hope you were paying attention as all of the PHY and MAC stuff I've been throwing at you will start to make sense in the upcoming chapters.

The gentleman in question from the last chapter was spending a bit of time in Canada as he decided he didn't want to finish out his time with the United States Navy. Being a musician at heart, Rick James formed the Mynah Birds with his buddy Neil Young. Little did anyone in the Mynah Birds know that the bands Steppenwolf and Buffalo Springfield would end up with a couple of the other Mynah Birds in their bands later on. Unlike The Byrds, which some think Rick got the Mynah Birds band name idea from, the black leather jackets, yellow turtlenecks and boots didn't make the Mynah Birds famous. Rick getting busted in the studio for being AWOL didn't help things much either. Needless to say, the Mynah Birds went by the wayside. However, the parts and pieces of the Mynah Birds produced and participated in big-name acts. If you can't hum "Heart of Gold" or sing to "Cinnamon Girl," go buy some Neil Young albums. Ever hear of Crosby, Stills, Nash and Young? Guess who Young is? Neil did some Buffalo Springfield time as well. Motown Records ended up with Rick James and old Ricky singlehandedly pulled Motown Records from the dust and back into contention with his 1981 hit "Super Freak." The Super Freak left us on August 7, 2004.

If you don't know who Bob Marley was, give this book to your sister, as she probably does know who Bob Marley was. I'll bet she also knows who was more popular in Jamaica than Bob Marley and had hits before Bob started jammin'. Do you?

A Look at the ZMD 900-MHz IEEE 802.15.4/ZigBee-Ready Radio

I sincerely hope you were wearing shoes while we were running through PHY and MAC hell. Your feet may be hot but you can now utilize the power of your mind to distance yourself from pain, as you are on the road to mastery of IEEE 802.15.4. Whether you like it or not, you are now a student of IEEE 802.15.4. There won't be any heavy IEEE 802.15.4 meditation sessions and I won't require you to strike a mid-air IEEE 802.15.4 Kung Fu pose while effortlessly defying gravity. While we're on Kung Fu movie stunts, the best one I ever saw had Kung Fu masters that could spin themselves into the ground and then cover themselves with dirt. These guys would spin into their holes along a path and as the bad guys approached they would spin themselves out of the ground and fight. Naturally, the spinning, hole-digging masters also had the ability to hang in the air forever while kicking the living crap out of their opponents. Once the fight was over, the masters would spin themselves back into hiding and await their next victims.

IEEE 802.15.4 Done the ZMD Way

The more you learn about how IEEE 802.15.4 works, the easier it will be to work with Zig-Bee. This book is all about doing IEEE 802.15.4 and ZigBee things. Right now, let's do some real IEEE 802.15.4 stuff. ZMD offers a really good IEEE 802.15.4 development kit that puts all of the elements of IEEE 802.15.4 together using ZMD44102 900-MHz IEEE 802.15.4 radio technology and basic ZMD44102 radio firmware. You've already been exposed to the ZMD44102 900-MHz transceivers, since I used them to check Beacon intervals in Chapter 3.

The ZMD44102 transceiver can operate in the 868-MHz band, which is the domain of European IEEE 802.15.4-compliant transceivers, and in the 915-MHz ISM in North America, Australia and New Zealand. I don't know about you, but I'm a cordless-phone freak. I like to keep up with the latest cordless-phone technology and I always send my Mom a new phone when I get one. Remember when cordless phones reached into the 900-MHz frequency band? Those new 900-MHz babies could talk farther and longer on a single battery charge than the existing 46/49-MHz cordless units. Plus, it was a little bit harder to eavesdrop on the 900-MHz phones, as a standard police scanner could tune you in to all of the neighborhood 46/49-MHz phones. Not too long after the 900-MHz phones became popular, 2.4-GHz phones showed up on the WalMart shelves. My take is that consumers saw the higher-frequency phones as technologically better since the frequency number was higher, and the marketing guys and gals knew that. It's kinda like buying a computer these days—the faster the better. My foray into cordless phones does have a point to make. The ZMD44102 900-MHz transceivers only compete with 900-MHz cordless phones and 900-MHz proprietary radio traffic. With the addition of 2.4-GHz cordless phones to the 2.4-GHz band, a 2.4-GHz IEEE 802.15.4-compliant transceiver now has to contend with Bluetooth traffic, 802.11b/802.11g WLAN traffic, proprietary radio traffic in the 2.4-GHz band and the ubiquitous microwave oven.

To further prove that 900 MHz is a viable frequency band for serious ZigBee work, the ZMD folks sniffed the air at the June 2006 ZigBee Open House. The results indicated that there was much more 2.4-GHz traffic than 900-MHz traffic competing for airspace.

The ZMD Starter Kit Bundle

The IEEE 802.15.4 concepts that will be offered up in this chapter will be brought to fruition using the Starter Kit Bundle version of the ZMD Wireless Sensor Starter Kit. That "bundle" means that our ZMD Wireless Sensor Starter Kit includes an extra ZMD44102 Starter Board that can be employed as a Daintree Networks SNA Sniffer capture module or an extra IEEE 802.15.4 node. The ZMD44102 Starter Board's ZMD44102 IEEE 802.15.4-compliant/ZigBee-ready single-chip 900-MHz transceiver serves under the command of a Silicon Laboratories C8051F120 microcontroller with an associated JTAG interface. Since the word "sensor" is synonymous with ZigBee and IEEE 802.15.4, the ZMD folks included one of their TSic 106 precision temperature sensors on each of the ZMD44102 Starter Boards. If you look just below and to the right of the Silicon Laboratories C8051F120 in Photo 4.1, you'll see the 3-wire TSic 106.

Photo 4.1: Silicon Laboratories C8051F120. The 10-pin header interfaces the Silicon Laboratories USB Debug Adapter to the ZMD44102 Starter Board's C8051F120. The ZMD44102 and all of its supporting electronics are directly to the right of the C8051F120. Power for the ZMD44102 Starter Board is supplied by the USB connection or the power jack. The switch bank to the right of the power jack configures the ZMD44102 Starter Board into either the 868-MHz or 900-MHz band and defines the ZMD44102 Starter Board as a central or remote node.

The Silicon Laboratories portion of the ZMD Wireless Sensor Starter Kit Bundle includes the C8051F120 microcontrollers, Silicon Laboratories' driver package for the Keil PK51 C compiler and a Silicon Laboratories USB Debug Adapter, which I've photographed in Photo 4.2.

Photo 4.2: Silicon Laboratories USB Debug Adapter. This little puppy provides a programming and debugging path between the Keil PK51 C compiler/μVision3 environment and the ZMD44102 Starter Board.

A 4K demo version of Keil's PK51 C compiler is included with the Starter Kit Bundle. However, you're going to need the full version of the Keil PK51 C compiler to fully utilize the supporting C source code that comes with the ZMD Wireless Sensor Starter Kit. For those of you that only want to evaluate the ZMD radio or the C8051F120 microcontroller, all of the C source code examples are also provided as ready-to-program hex files that you can push down into the C8051F120 using the USB Debug Adapter and Silicon Laboratories Flash programming utility that come with the ZMD Starter Kit Bundle.

The ZMD Starter Kit Bundle also comes with a 30-day trial version of Daintree Networks SNA. The Daintree Networks SNA (Sensor Network Analyzer) package is a professional IEEE 802.15.4 packet sniffer. Everything you need to know about a transmitted IEEE 802.15.4 packet is available via the Daintree Networks SNA application's integral window panes. Fortunately, the Daintree Networks SNA demo included with the ZMD Wireless Sensor Starter Kit is the professional version, which allows you not only to see the bits, but to visualize a ZigBee-enabled IEEE 802.15.4 network as well. The ZMD44102 Starter Boards are compatible with Daintree Networks SNA. So, we can and will use one of the ZMD44102 Starter Boards as the Daintree Networks SNA 900-MHz capture device.

The ZMD44102 Transceiver

The ZMD44102 CMOS transceiver you're eyeballing in Photo 4.3 is an 868.3-MHz/ 902-928-MHz band single-chip multichannel IEEE 802.15.4-compatible system-on-chip device. License-free operation is provided by the ZMD44102 on the 868.3-MHz European band and the 902-MHz-to-938-MHz North American ISM (Industrial, Scientific and Medical) bands. The ZMD44102 is fully capable of obtaining the maximum burst data rate of 20 Kbps in the

868-MHz band. It can also reach the maximum 900-MHz band data rate of 40 Kbps. The ZMD44102 employs the services of Direct Sequence Spread Spectrum technology (DSSS), which provides a reliable means of data transfer in high-traffic and hostile RF environments. Keeping with the simplicity associated with IEEE 802.15.4 devices, the ZMD44102 is highly integrated and requires a minimum of external components for proper operation. ZMD44102 line-of-sight transmission distances can reach beyond 100 meters.

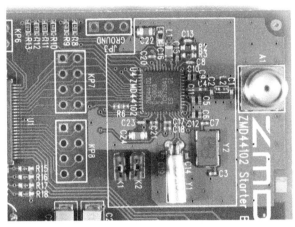

Photo 4.3: ZMD44102 CMOS transceiver. It looks just like its 2.4-GHz cousins but it doesn't have to fight with the microwave oven for air time.

The ZMD44102 CMOS transceiver you're eyeballing in Photo 4.3 is an 868.3-MHz/ 902-928-MHz band single-chip multichannel IEEE 802.15.4-compatible system-on-chip device. License-free operation is provided by the ZMD44102 on the 868.3-MHz European band and the 902-MHz-to-938-MHz North American ISM (Industrial, Scientific and Medical) bands. The ZMD44102 is fully capable of obtaining the maximum burst data rate of 20 Kbps in the 868-MHz band. It can also reach the maximum 900-MHz band data rate of 40 Kbps. The ZMD44102 employs the services of Direct Sequence Spread Spectrum technology (DSSS), which provides a reliable means of data transfer in high-traffic and hostile RF environments. Keeping with the simplicity associated with IEEE 802.15.4 devices, the ZMD44102 is highly integrated and requires a minimum of external components for proper operation. ZMD44102 line-of-sight transmission distances can reach beyond 100 meters.

The low-power 20/40-Kbps ZMD44102 transceiver is also equipped with a complete IEEE 802.15.4–compliant PHY and a thin MAC that is implemented in hardware. The ZMD44102 MAC houses a 128-byte transmit FIFO (First In First Out) buffer and a 256-byte receive FIFO. That's just enough temporary buffer area for one IEEE 802.15.4 frame going out and two incoming IEEE 802.15.4 frames. Even though the ZMD44102 MAC can buffer up a couple of incoming IEEE 802.15.4 frames, we will want to clear the input FIFO as quickly as possible to prevent the loss of data. What good would any MAC be if it could not automatically check and generate a CRC (Cyclic Redundancy Check)? The ZMD44102's MAC doesn't have to worry about being substandard, as it generates frame CRCs and checks incoming frames for the correct CRC.

Remember slotted and unslotted networks? I told you there would be a test later. The ZMD44102 can be used in standard unslotted CSMA-CA networks and in slotted CSMA-CA networks. When used in slotted networks, the ZMD44102 has the capability to handle the GTS area that may be associated with a superframe.

In addition to its ability to easily operate in slotted and unslotted network situations, the ZMD44102 MAC operates in a multitude of other modes. The ZMD44102 MAC has to receive as well as transmit. If the ZMD44102 MAC belongs to a PAN Coordinator, the ZMD44102 MAC will be asked to perform active scans to determine RF energy levels (ED) or find the most suitable channel on which to start a new PAN. If the ZMD44102 MAC finds itself in the End-Device role, a passive scan may be requested in an attempt to find a Beacon that would signal a PAN Coordinator is within range. It's the MAC's job to execute that scan. Beacon generation and Beacon tracking are both tasks that are handled by the ZMD44102 MAC, depending on whether the device associated with the ZMD44102 MAC is a PAN Coordinator or an End Device.

Many of the MAC tasks such as scanning and superframe component handling are regulated by time. Normally, peripheral interrupts and microcontroller timers would be put on the front line to handle the timekeeping and task scheduling. Servicing interrupts and shepherding internal timers gobble up microcontroller resources. To help keep the microcontroller focused on the application instead of the ZMD44102 MAC's timing requirements, the ZMD44102 incorporates a gaggle of its own internal timers. The ZMD44102 uses a pair of separate clocks to pull off its timing tricks. The 24-MHz system clock is used to clock the digital core and a separate 32.768-kHz clock, which supports sleep and power-down modes, and acts as an RTC. Both of the ZMD44102's internal clocks, in one manner or another, support the set of IEEE 802.15.4-oriented timers that are integrated into the ZMD44102's HW-MAC. The RTC continues to run when the 24-MHz system clock is shut down during sleep and Global Power Down modes. Shutting down the 24-MHz clock reduces the ZMD44102's power consumption. Since the RTC runs full time, it can maintain network time. This allows the ZMD44102 in an End Device to sleep and always keep track of superframe time. Recall that in slotted networks the best thing for an End Device to do is sleep and wake just before the superframe that will service it begins.

Most IEEE 802.15.4/ZigBee transceivers use an SPI portal to communicate with a host microcontroller. By implementing an industry-standard SPI portal, the ZMD44102 allows the IEEE 802.15.4 designer to pair the ZMD44102 up with just about any of today's off-the-shelf microcontrollers. There may be instances when the SPI hardware interface is not accessible. No worries. The ZMD44102 is also capable of passing data and commands over a parallel interface made up of the host microcontroller's general-purpose I/O pins.

The ZMD Wireless Sensor Starter Kit documentation uses the IEEE 802.15.4 standard's nomenclature in their datasheet and user manual language. That's a good thing. As you read about the ZMD44102, you can directly relate its operation to the relevant sections of the IEEE 802.15.4 standard.

Preflighting the ZMD44102

I thoroughly enjoyed getting the ZMD Starter Kit Bundle up and running. The demo application code and Silicon Laboratories drivers that came with my ZMD Wireless Sensor Starter Kit were designed to be used with Keil's uVision2 environment. We will be using the latest version of the Keil PK51 C compiler, which is built around uVision3. After some investigation, or should I say, after things wouldn't work quite right, I found that the Silicon Laboratories USB Debug Adapter required the uVision3 drivers loaded in order to enable the recognition of the USB Debug Adapter's USB interface from within the Keil PK51 C compiler's debug utilities set-up windows. Once I got the Keil PK51 C compiler to recognize the Silicon Laboratories USB Debug Adapter, I performed a test compile of the BCS (Basic Communication Software) source that came with the ZMD Wireless Sensor Starter Kit. Everything came out lovely and I engaged the uVision3's virtual DEBUG button to pour my newly compiled BCS hex file into the ZMD44102 Starter Board's C8051F120 Flash. The idea behind the ZMD Starter Kit Bundle is to allow the development kit user to apply the ZMD44102 hardware directly to the mechanisms of IEEE 802.15.4. The actual "application" stuff can then be layered on top of the transport foundation the ZMD44102 Starter Boards provide. A ZigBee or IEEE 802.15.4 application does not have to be complex. The simple act of reading a temperature or monitoring a switch is considered an application in the ZigBee and IEEE 802.15.4 worlds.

I want to continue to show you what is going out of the network node antennae. Since the ZMD44102 is a 900-MHz radio, we must configure the Daintree Networks SNA for 900-MHz band operation and prepare the associated Daintree Networks SNA IEEE 802.15.4-compliant frame-capture hardware. It's only logical (and necessary) to use the third ZMD44102 Starter Board as the frame-capture hardware for the Daintree Networks SNA. The ZMD44102 Starter Board is compatible with the Daintree Networks SNA application. An application note on the Daintree web site provides in detail what is required to put a ZMD44102 to work as a Daintree Networks SNA capture device. In addition, a precompiled and ready-to-load Daintree driver hex file is included with the ZMD Wireless Sensor Starter Kit Bundle. All we have to do is load up the Silicon Laboratories standalone Flash programming utility, which also was a component of the ZMD Wireless Sensor Starter Kit Bundle, and drop the Daintree Networks SNA driver/capture code into the C8051F120 Flash on the newly tasked ZMD44102-based IEEE 802.15.4 capture module.

Firing Up the ZMD44102

Before we jump off into turning on radios, there's business that the microcontroller needs for us to take care of first. The initialization of the ZMD44102 Starter Board is taken care of up front in the *main* function of the BCS. In this case, the C8051F120 is the target microcontroller and the function calls in Code Snippet 4.1 invoke the instructions needed to get the C8051F120 on course.

Code Snippet 4.1

```
**********************************************************************
   WD_DISABLE();
   SYSCLK_Init();             /* initializes system clock */
   portInit();                /* initializes crossbar, i/o lines and interrupts */
   zmdInitHal();              /* initializes the ZMD's HAL */
   mlmeResetRequest(TRUE);    /* init the PHY and MAC layer with its default
                                 values */
   EA = 1;                    /* enable global interrupts */
   zmdGpd(OFF);
   zmdReset(HOLD_TIME_3);     /* resets the ZMD44102 */
**********************************************************************
```

Code Snippet 4.1: This is standard stuff as far as microcontrollers go. The Silicon Laboratories C8051F120 gets its clock and general-purpose I/O configuration from the SYSCLK_Init, portInit and zmdInitHal functions.

The SYSCLK_Init function essentially only consists of a single instruction that commands the C8051F120 to use its internal 24.5-MHz oscillator as its clock source. The C8051F120 uses a digital crossbar to multiplex digital peripherals and I/O ports to selected general-purpose I/O pins. The code within the portInit function works on the C8051F120's crossbar and sets up the desired general-purpose I/O configuration.

We discussed the HAL earlier. If you're thinking that the coding of the HAL is based on the C8051F120 interfacing to the ZMD44102, you're correct. The ZMD44102 HAL lays the groundwork for the zmdGpd (ZMD44102 Global Power Down) and zmdReset functions listed in Code Snippet 4.1. The zmdReset timing controls the mode that the ZMD44102 will reset into. For instance, in Code Snippet 4.1, the ZMD44102 reset argument specifies a RSN pin hold time of 730 ms, which will reset into the Off mode. If the zmdReset argument were HOLD_TIME_1 (500 µs), the ZMD44102 would reset into idle mode. Sleep or Global Power Down reset mode is entered when the zmdReset argument is HOLD_TIME_2 (3 ms). C8051F120 SPI initialization and millisecond/microsecond timing routines are also defined within the ZMD HAL code. SPI support is future enhanced within the HAL functions with calls for SPI byte reads and writes as well as SPI block read and write routines.

Does MLME ring a bell? BONG!! MLME is short for MAC Layer Management Entity. The leading "mlme" in the mlmeResetRequest function call tells us that a MAC management action is about to be requested. The code we are looking at in Code Snippet 4.1 is technically one layer above the MAC sublayer. The semantics of the mlmeResetRequest function call follow the service primitive semantics called out in the IEEE 802.15.4 standard. So, technically the MLME-RESET.request primitive has been put into motion via the MLME-SAP. What should happen as a result of the mlmeResetRequest function call's TRUE argument is that the MAC sublayer gets reset and all of the MAC PIB attributes are reset to their default values. The ZMD implementation of the mlmeResetRequest primitive issuance also resets the PHY PIB attributes to their defaults. Every MAC PIB and PHY PIB attribute that matters to the ZMD44102 is set to defaults in the code contained within Code Snippet 4.2.

Code Snippet 4.2

```
****************************************************************************
void mlmeResetRequest(BOOL SetdefaultPIB)
{
    if (SetdefaultPIB)
    {
        phyPIBInit(TRUE);    // init phy layer with default values
            macPIB.macBeaconPayloadLength = 0;
            macPIB.macBeaconOrder = 0x0F;
            macPIB.macBSN = 0x00;
            macPIB.macCoordShortAddress = 0xFFFF;
            macPIB.macDSN = 0x00;
            macPIB.macPANId = 0xFFFF;           // not associated
            macPIB.macShortAddress = 0xFFFF; // no short address
            macPIB.macSuperframeOrder = 0x0F;
    }
}

void phyPIBInit (BOOL defaultValue)
{
    if (defaultValue)
    {
        phyPIB.phyCurrentChannel = 0x00;
        phyPIB.phyChannelsSupported = 0x7FF;
        phyPIB.phyTransmitPower = ZMD_TX_PWR_0;
        phyPIB.phyCCAMode = CCA_THRESHOLD;
    }
}
****************************************************************************
```

Code Snippet 4.2: This is a ZMD implementation of the MLME-RESET.request primitive. The ZMD coders decided that it was a good idea to go ahead and initialize the PHY too. Note that the default MAC Beacon Order and Superframe Order values define an unslotted network.

OK. The C8051F120 crossbar and general-purpose I/O are configured correctly and the ZMD44102 has a viable SPI connection to the C8051F120. The ZMD44102 is sitting in a default mode as we have only issued a power-on reset to the device and set its GPD pin to an inactive state, preventing the ZMD44102 from entering a Global Power Down state. So far, we have not spoken with the ZMD44102 via the C8051F120's SPI portal.

Our First Steps

We are about to leave our crawling stage and put a step or two together with our first IEEE 802.15.4 data transmission. I didn't mention the words network or ZigBee because we won't be establishing or joining a network and we won't be utilizing any of ZigBee's sublayer services to send our little three-letter message. In fact, we don't even have an End Device configured to listen to us other than the Daintree Networks SNA capture device. Since there is no official ZigBee or IEEE 802.15.4 network, the visualization feature of the Daintree Networks SNA won't be able to draw us up a pretty picture of the network. That's OK. The Daintree Networks SNA will still capture whatever bits we are able to throw out of the ZMD44102 Starter Board's antenna.

The code shown in Code Snippet 4.3 is what we will use to send the characters "ZMD" over channel 1 of the 900-MHz ISM band.

Code Snippet 4.3

```
****************************************************************************
void UnslottedTx (void)
{
    unsigned char ack;
    unsigned char channel;
    unsigned char tx_data[125];
    unsigned char payload_length;
    unsigned int destpanid, destaddr;

    destpanid = 0xBEEF;
    destaddr = 0xFEED;
    channel = 0x01;
    ack =0x01;
    tx_data[0] = 'Z';
tx_data[1] = 'M';
tx_data[2] = 'D';
    payload_length = 0x03;
    plmeSetRequest(phyCurrentChannel, &channel);
    zmdUnslottedTx(ack, tx_data, payload_length, destpanid, destaddr,
        AM_SHORT_ADDR);
    }
}
****************************************************************************
```

Code Snippet 4.3: Do you recognize the plme prefix? PHY Layer Management Entity should have rolled right off your tongue. It looks like a PHY management primitive is being used to request the ZMD44102 to switch to and use channel 1. What do you think?

The functions and services provided by the ZMD BCS code package functions, such as the one you see in Code Snippet 4.4, are responsible for making the little application in Code Snippet 4.3 so simple to conceive. In the IEEE 802.15.4 specification, the PHY is defined with integral data (PD-SAP) and management (PLME-SAP) SAPs (Service Access Points), which are logical portals that pass primitives between the PHY and MAC sublayers. Since the PHY's PLME-SAP and PD-SAP are the only SAPs that join the MAC and PHY sublayers, the MAC will always pass the final management or data primitive to the PHY sublayer and any data the PHY needs to pass up the stack will always flow through the MAC. Recall that a primitive in the true IEEE 802.15.4 sense does not within itself contain any executable elements or elements that will directly invoke program execution. A primitive simply carries what is needed to assist in the action that the primitive is intended to initiate. In reality, primitives are simply data structures that contain information that can be used and built upon by the neighbor sublayer the primitive is passed to. Instead of passing the PLME-SET.request primitive in the traditional IEEE 802.15.4 manner, the ZMD BCS code calls an executable function using the primitive name. The "plme" portion of the plmeSetRequest function stands for PHY Layer Management Entity. The PLME, as it is called in the IEEE 802.15.4 spec, is responsible for coordinating any management functions associated with the PHY. If a PLME-SET.request primitive were issued in the IEEE 802.15.4 by-the-book method, a PLME-SET.confirm primitive would normally be expected in response. The ZMD plmeSetRequest function returns the status directly to the function caller in lieu of passing a formal confirmation primitive. If you understand the idea behind primitives, the ZMD IEEE 802.15.4-like functions should not ruffle your feathers. The ZMD IEEE 802.15.4 shortcut methods work very well and the final results of executing a ZMD BCS function instead of passing a primitive in the traditional way are identical.

Code Snippet 4.4
****

```
 PHY_ENUM plmeSetRequest (PHY_PIB_ATTR PIBAttribute, void
*PIBAttributeValue)
{
    switch (PIBAttribute)
    {
        case phyCurrentChannel:
            if (*(UINT8*)PIBAttributeValue > 10) return PHY_INVALID_PARAMETER;
            phyPIB.phyCurrentChannel = *(UINT8*)PIBAttributeValue;
            zmdWriteReg(PHY_CHANNEL, phyPIB.phyCurrentChannel);
            break;
        case phyChannelsSupported: // this is a read only parameter
            return PHY_INVALID_PARAMETER;
        case phyTransmitPower:
            if (*(UINT8*)PIBAttributeValue > 3) return PHY_INVALID_PARAMETER;
            phyPIB.phyTransmitPower = *(UINT8*)PIBAttributeValue;
            {
                UINT8 temp_value;

temp_value = (zmdReadReg(TX_MODE) & 0xCF) | (phyPIB.phyTransmitPower << 4);
                zmdWriteReg(TX_MODE, temp_value);
            }
            break;
        case phyCCAMode:    // this is a read only parameter
            return PHY_INVALID_PARAMETER;

        default: return PHY_UNSUPPORTED_ATTRIBUTE;
    }
    return PHY_SUCCESS;
}
```

****

Code Snippet 4.4: This ZMD version of a PLME-Set.request primitive services four PHY PIB attributes. Note that the channels supported and CCA Mode PHY PIB attribute values cannot be altered in the ZMD BCS package. However, the ZMD BCS code is so easy to follow, you can alter the BCS code to change those read-only PHY PIB attributes to alterable variables.

The important thing to come away with from the code in Code Snippet 4.4 is that, once the attribute value is placed into its position within the PHY PIB, the zmdWriteReg function launches the bits across the C8051F120-to-ZMD44102 SPI portal into the associated ZMD44102 transceiver register.

Code Snippet 4.5 takes all of the PAN, attribute and data values we specified in Code Snippet 4.3 and places them into the corresponding ZMD44102 registers. Let's follow the flow of the code in Code Snippet 4.5.

The ZMD44102 is reset into idle mode and the ZMD44102's threshold trimming and register values are set to their optimum values by the code contained within the zmdInit function. A plmeGetRequest function call requesting the current channel is issued within zmdSetPibValues. If the PHY returns the channel number successfully, the current channel is loaded into the ZMD44102. A second plmeGetRequest function call issued within zmdSetPibValues retrieves the transmit power setting, which is also transferred via a zmdWriteReg SPI transaction to the proper ZMD44102 register.

Code Snippet 4.5
```
****************************************************************************
MAC_STATUS_ENUM zmdUnslottedTx (BOOL ack, UBYTE *payload, UINT8 payload_
length, UINT16 dest_pan_id, UINT16 dest_addr16, ADDR_MODE addr_mode)
{
    UBYTE i;
    UBYTE buffer[aMaxPHYPacketSize];
    UBYTE irq_reason, rx_status, tx_status;

    zmdReset(HOLD_TIME_1);
    zmdInit();
    zmdSetPibValues();
/* set the transmit frame MAC payload length */

    zmdWriteReg(MSDU_TX_LENGTH, payload_length);
/* set sequence no */
    zmdWriteReg(MHR_TX_SEQ_NO, macPIB.macDSN++);
    if (addr_mode == AM_SHORT_ADDR)
    {
        /* set pan id */
zmdWriteReg(MHR_TX_DST_PAN_ID_1, LOW_BYTE(dest_pan_id));
        zmdWriteReg(MHR_TX_DST_PAN_ID_2, HIGH_BYTE(dest_pan_id));
        /* set dest addr */
        zmdWriteReg(MHR_TX_DST_ADDR16_1, LOW_BYTE(dest_addr16));
        zmdWriteReg(MHR_TX_DST_ADDR16_2, HIGH_BYTE(dest_addr16));
        /* set source addr */
        zmdWriteReg(MHR_TX_SRC_ADDR16_1, LOW_BYTE(macPIB.macShortAddress));

        zmdWriteReg(MHR_TX_SRC_ADDR16_2, HIGH_BYTE(macPIB.macShortAddress));
        /* set the transmit frame MAC header frame control field, higher bits */
        zmdWriteReg(MHR_TX_FC_2, FT_DST_ADDR_MODE_16 | FT_SRC_ADDR_MODE_16);
    }
    /* configure tx: use CSMA */
zmdWriteReg(MAC_TX_CONFIG, MC_CSMA);
    if (ack == ACK_REQUEST)
    {
/* set the transmit frame MAC header frame control field, using Ack */
        zmdWriteReg(MHR_TX_FC_1, FT_DATA | FT_ACK_REQUEST | FT_INTRA_PAN);
```

```
/* store ACK and LQI, auto check sequence number enabled */
zmdWriteReg(MAC_RX_CONFIG, MC_FIFO_STORE_ACK | MC_FIFO_STORE_LQI | MC_ACK_
SQU_NB_CHECK_EN);
    }
    else
    {
/* set frame type */
        zmdWriteReg(MHR_TX_FC_1, FT_DATA);
    }
/* write payload data to tx fifo */
    zmdWriteTxFifo(payload_length, payload);
/* activate transmission */
    zmdWriteReg(MAC_CTRL, MC_TX_ON);
```
****

Code Snippet 4.5: Just plug in the matching stuff from Code Snippet 4.3 and this should all make sense to you. We are simply loading up the appropriate ZMD44102 registers and turning on the transmitter.

As you follow down through the code progression in Code Snippet 4.5, note that only the three bytes of payload data are actually loaded into the ZMD44102's transmit FIFO. The ZMD44102 contains an internal frame-forming engine that automatically plucks the required frame elements from the ZMD44102 registers and melds them with the payload data for transmission. If the programmer desires to have complete control of the frame-assembly process, the ZMD44102's frame-forming automation can be programmatically disabled.

Executing our little UnslottedTX function resulted in the transmission of fourteen bytes, which were all captured by our ZMD44102 Starter Board turned Daintree Networks SNA 900-MHz capture module. The Daintree Networks SNA packet decode is shown in Screen Capture 4.1.

Screen Capture 4.1: Again, just match up the familiar items with what we've been discussing and that light bulb hanging above your noggin should illuminate.

Let's analyze the contents of Screen Capture 4.1. The actual frame-length calculation, which is gathered from the PPDU (PHY Protocol Data Unit) header area, is correct. Including the pair of CRC bytes that are shown as dots at the end of the hex dump, I count 14 octets in the hex dump area of Screen Capture 4.1.

The Frame Control Word resides inside of the MAC protocol data unit (MPDU) header, which is identified by MHR (MAC header) in the IEEE 802.15.4 specification. I added the FT_IN-TRA_PAN bit to the low byte of the Frame Control Header in Code Snippet 4.5 as there will be no intra-PAN communications in our simple little IEEE 802.15.4 network. My code change is reflected by the Intra PAN bit value shown in Capture 4.1. Turning on the intra-PAN bit in the Frame Control word eliminates two bytes (Source PAN Identifier) from the addressing fields.

The broken-down Frame Control word tells us that the Destination Addressing and Source Addressing Modes are 16 bits in length, which means that 16-bit short addresses will be used instead of the 64-bit IEEE addresses. Most every IEEE 802.15.4 and ZigBee implementation uses the 16-bit addressing modes rather than the IEEE 64-bit addressing modes as the less stuff you're sending, the more power you're saving. Do the Destination PAN Identifier and the Destination Address values look familiar? How about the Source Address value? According to Code Snippet 4.2, the source Address should be 0xFFFF as macPIB.macShortAddress = 0xFFFF is defined. Well, before I compiled the UnslottedTx application, I changed the macPIB.macShort-Address value in the MAC PIB to read ASCII "SA" for Source Address. The MAC PIB (PAN Information Base) is kept inside the mlmeResetRequest function in the BCS code. If you follow the IEEE 802.15.4 rule book, which the BCS is doing in its own way, the logical location of the MAC PIB should be inside of the MLME (MAC Sublayer Management Entity), which is part of the MAC sublayer.

It looks like it took me fourteen tries at sending and capturing this frame (it did), as the Sequence Number is greater than one. Even though we asked for an acknowledgment to be sent upon reception of this frame, there was no End Device to receive our data and, thus, an acknowledgment from a nonexistent End Device cannot be generated or transmitted.

The frame captured in Capture 4.1 is about as raw and basic as it gets. There is absolutely nothing ZigBee in this gaggle of bits at all. Everything you see in Screen Capture 4.1 is IEEE 802.15.4 PHY and MAC related. All we did here was stuff the Frame Control bits, make up some addressing words, stick a text message behind it all and cram it into the ZMD44102's registers and transmit FIFO. The ZMD44102 added the necessary frame encapsulation pieces (preamble, SFD (Start of Frame Delimiter), CRC), formed up the IEEE 802.15.4 frame and pushed the whole mess out the ZMD44102 Starter Board's antenna.

No rules were broken and there is absolutely nothing wrong with the frame we just aired. As long as the receiver and the receiving application know what to do with the incoming frame, who cares about its format? For instance, the receiver could have simply parsed or counted through the Frame Control and addressing fields of the IEEE 802.15.4 frame we sent and picked out what it wanted to use of the text message. Or, if there were more than one receiver, the address information could have also been parsed and logically analyzed to determine who the message was really intended for. That's essentially what a ZigBee stack does, but it does it quite a bit more elegantly.

Our First Network…Sorta

Let's be a bit more elegant as well. I say we build up a simple two-node unslotted network and send a meaningful byte of data over it. Recall that the term "unslotted" means that the little ad hoc IEEE 802.15.4 network we're about to build will not use Beaconing, which means there will be no special, exclusive or repetitive time slots allocated for data transfer timed to a recurring Beacon signal. In other words, no superframes and no GTSs. If the coast is clear, fire when ready.

All we'll really need to do is to determine if we want an acknowledgment of the transmission, which we do, identify some specific transmitter and receiver PAN and node addresses, which we have, specify the length of the data we wish to send—did that too—and plug in our data. You've already seen that the ZMD folks have done most of the work for us already in the BCS code. Once we have nailed down who, what and where, all we have to do is call upon some of the functions included within the BCS. We will need another ZMD44102 Starter Board to act as a receiver. That means we will also need some receiver code. The ZMD44102 Starter Board receiver node code is notated in Code Snippet 4.6.

Code Snippet 4.6
```
****************************************************************************
void UnslottedRx (void)
{
    UINT8 channel;
    UBYTE data_buffer[aMaxPHYPacketSize];
    UBYTE i,j;

    channel = 0x01;
    plmeSetRequest(phyCurrentChannel, &channel);
    while (1)
    {
    if (RX_FAILED == zmdUnslottedRx(&data_buffer, TRUE))
     break;
    else
    {
     wait_ms(5);
     for(j=0;j<20;j++)
     i = data_buffer[j];
     P3 = data_buffer[10];
    }
}
}
****************************************************************************
```
Code Snippet 4.6: There's not much you can't get your arms around here. A good guess for the value of the aMaxPHYPacketSize variable is 127. What do you think?

The Boolean variable auto_ack in Code Snippet 4.7 is indeed TRUE as things get turned over quickly to the zmdUnslottedRX function. Thus, the very first thing that is done to the receive configuration is to turn on the auto acknowledgment bit and allow the LQI (Link Quality Indication) word to be stored behind the packet in the ZMD44102's receive FIFO. Then, the ZMD44102's transmit success interrupt trigger is disabled.

Code Snippet 4.7

```
*****************************************************************************
MAC_STATUS_ENUM zmdUnslottedRx (UBYTE *data_buffer, BOOL auto_ack)
{
    UBYTE rx_status, irq_reason;
    zmdReset(HOLD_TIME_1);                              zmdInit();

    zmdSetPibValues();
/* enable frame filter: frames with CRC failure are ignored */
    zmdWriteReg(MAC_FILT_CONFIG, MFC_LVL1_FILT_EN);
    if (auto_ack == TRUE)
    {
        zmdWriteReg(MAC_RX_CONFIG, MC_AUTO_ACK_EN | MC_FIFO_STORE_LQI);
        /* no interest in ack sending status */
zmdWriteReg(IRQ_MASK_2, MASK_TX_SUCCESS);
}
    else
        zmdWriteReg(MAC_RX_CONFIG, MC_FIFO_STORE_LQI);

/* switch the receiver on */
    zmdWriteReg(MAC_CTRL, MC_RX_ON);
    /* wait for the IRQ and get its reason, blocking function */
irq_reason = zmdWaitForIRQ();
    /* get the receive status */
    rx_status = zmdReadReg(MAC_RX_STATUS);
    /* check the receive status */
    if ((rx_status == MS_RX_DATA) || (rx_status == MS_RX_DATA_ACK))
    {
        /* get the received data from the rx fifo and stored it to the data_
            buffer */
        if (zmdGetRxPacket(data_buffer))            {
        /* received data is in the data_buffer; leave receive function */
        return SUCCESS;
    }
    else
    {
        zmdWriteReg(MAC_CTRL, MC_TRX_OFF);
        return RX_FAILED;
    }

    zmdWriteReg(MAC_CTRL, MC_TRX_OFF);
    return SUCCESS;
}
*****************************************************************************
```

Code Snippet 4.7: Pay special attention to the frame filtering. We are able to send and receive without regard to addressing, as the MAC is only examining the CRC of each incoming frame.

The receiver gets activated and as soon as some valid data (good packet CRC) gets through to the receive FIFO, control is returned to the UnslottedRX function. The 5mS wait is there to allow time for the link to turn around and the receiver MAC to post and transmit an acknowledgment message. I allowed 5 ms here but it really only takes about 3 ms for this to happen. The zmdGetRxPacket function has already transferred the data from the receive FIFO to the data_buffer array. So, all we have to do is parse through the data_buffer array and do with its contents as we please. I know that the actual data payload is at offset 0x0A as there is a packet length byte at offset 0x00 of the data_buffer array. Just to make something physical happen, I dump the payload data byte into the receiving ZMD44102 Starter Board's LED I/O port. You can see the contents of the data_buffer array after the reception of the packet from our transmitting node in Screen Capture 4.2.

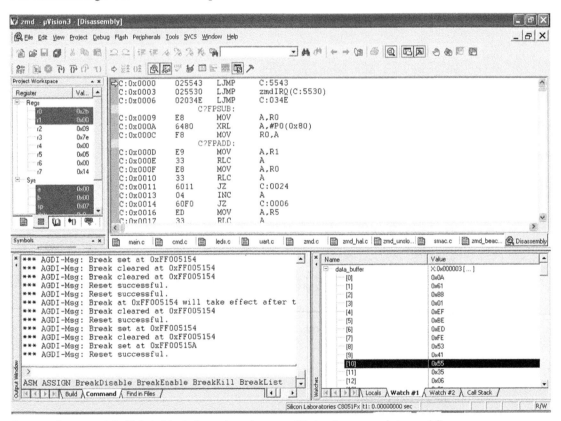

Screen Capture 4.2: This is kinda busy but the proof is in the array data pudding.

Take a look at the Sequence Number in Screen Capture 4.3. It matches up with the Sequence Number in Screen Capture 4.2 and the sequence number in Screen Capture 4.4. We successfully twisted all of the necessary buttons and knobs (the acknowledge request bit set in the Frame Control word and the AutoAckEnable bit set in the receive configuration) to allow the ZMD44102's MAC to generate the 5-byte acknowledge message shown in Screen Capture 4.3.

```
Packet Decode                                                                    _ □ ✕

Frame 2 (Length = 5 bytes)
   Time Stamp: 15:38:03.949
   Frame Length: 5 bytes
   Capture Length: 5 bytes
   Link Quality Indication: 231
IEEE 802.15.4
   Frame Control: 0x0002
      .... .... .... .010  = Frame Type: Acknowledgment (0x0002)
      .... .... .... 0...  = Security Enabled: Disabled
      .... .... ...0 ....  = Frame Pending: No more data
      .... .... ..0. ....  = Acknowledgment Request: Acknowledgement not required
      .... .... .0.. ....  = Intra PAN: Not within the PAN
      .... ..00 0... ....  = Reserved
      .... 00.. .... ....  = Destination Addressing Mode: PAN identifier and address field are not present (0x0000)
      ..00 .... .... ....  = Reserved
      00.. .... .... ....  = Source Addressing Mode: PAN identifier and address field are not present (0x0000)
   Sequence Number: 1
   Frame Check Sequence: Correct

0000:   02 00 01 .. ..                                            ..|..
```

Screen Capture 4.3: This acknowledgment message is automatically generated by the ZMD44102 MAC in response to the data frame received from the transmitting node in our dual-node IEEE 802.15.4 network.

I hope you're wondering why all of the PAN, destination and source addresses are the same and everything works just fine. Well, all we really have here is a simple peer-to-peer network. There are no official PAN Coordinators or End Devices and there is no ZigBee NWK layer to assist in the formation and operation of our little juvenile delinquent network. The need for address management was eliminated by this line of code from Code Snippet 4.7:

```
*********************************************************************************
/* enable frame filter: frames with CRC failure are ignored */
    zmdWriteReg(MAC_FILT_CONFIG, MFC_LVL1_FILT_EN);
*********************************************************************************
```

There are three levels of MAC filtering supported by the ZMD44102. Level 1, which is invoked by the code I just showed to you, only checks the incoming frames for a good CRC. Therefore, any IEEE 802.15.4 frame with the correct CRC will be allowed into the ZMD44102's receive FIFO. Level 2 filtering is not enabled and if it were the frame type, address and PAN identifier would be scrutinized. The functions provided by Level 3 frame filtering are as follows:

- Reject all nonacknowledge frames while waiting for an acknowledgment

- Reject all nonBeacon frames in Beacon track mode during the Beacon scan phase

- Reject all nonBeacon frames during active and passive scan

- Reject all noncommand frames during orphan scan

By eliminating frame filtering at Levels 2 and 3, we put the ZMD44102 MAC into semipromiscuous mode.

```
Packet Decode                                                        _ □ X
Frame 1 (Length = 12 bytes)
   Time Stamp: 15:38:03.947
   Frame Length: 12 bytes
   Capture Length: 12 bytes
   Link Quality Indication: 248
IEEE 802.15.4
   Frame Control: 0x8861
      .... .... .... .001  = Frame Type: Data (0x0001)
      .... .... .... 0...  = Security Enabled: Disabled
      .... .... ...0 ....  = Frame Pending: No more data
      .... .... ..1. ....  = Acknowledgment Request: Acknowledgement required
      .... .... .1.. ....  = Intra PAN: Within the PAN
      .... ..00 0... ....  = Reserved
      .... 10.. .... ....  = Destination Addressing Mode: Address field contains a 16-bit short address (0x0002)
      ..00 .... .... ....  = Reserved
      10.. .... .... ....  = Source Addressing Mode: Address field contains a 16-bit short address (0x0002)
   Sequence Number: 1
   Destination PAN Identifier: 0xbeef
   Destination Address: 0xfeed
   Source Address: 0x4153
   Frame Check Sequence: Correct
   MAC Payload: 55

0000:   61 88 01 ef be ed fe 53 41 55 .. ..         a..o>m~SAU..
```

Screen Capture 4.4: Out of all of the twelve bytes sent, only one byte is of importance to us.

The lone MAC payload byte (0x55) shown in Screen Capture 4.4 is singled out and written to the ZMD44102 Starter Board's LED general purpose I/O pins.

We're On Our Way

We've taken our first couple of steps towards walking in ZigBee and IEEE 802.15.4 network land. Be sure to get very comfortable with all of the IEEE 802.15.4 concepts we've covered thus far, because we're going to step it up a notch in the next chapter.

About ZMD

ZMD was founded in 1961 and over the past 45 years it has played a significant role in the rapid development of the microelectronics industry. The company is said to be the cradle of the Saxon microelectronics industry and is a founder of the largest European semiconductors cluster, "Silicon Saxony". That's nice. My experience with ZMD has been through a relationship with William Craig, who supplied the ZMD content for this book. Here's yet another human that cares about you if you care about 900-MHz IEEE 802.15.4 radios. Bill actually returns phone calls and responds to emails. Thanks, Bill. You can find Bill's moniker amongst the ZigBee Alliance elite.

By the way, the reggae king that paved the way for Bob Marley was Desmond Dekker. Desmond was into ska as well. If you don't know what ska is, try listening to some No Doubt music, as that Gwen Stefani-fronted band is heavily influenced by ska. Desmond's big hit was done with his backing band, The Aces. The song was called Israelites and was released in 1969. I was to become an AM radio DJ the following year. I played the hell out of Desmond's Israelites on my nightly show. Desmond left us on May 25, 2006.

Here's an easy one for you. Who is considered the Guy Lombardo of Halloween?

Atmel Does IEEE 802.15.4 and ZigBee Too

Atmel does ZigBee and IEEE 802.15.4 too. In this chapter we'll explore ZigBee and IEEE 802.15.4 from an Atmel perspective. I have five Atmel Z-Link radio modules that we can have some fun with. One of them agreed to a photo op and posed for Photo 5.1. So, let's open the ZigBee/IEEE 802.15.4 door that says "Atmel."

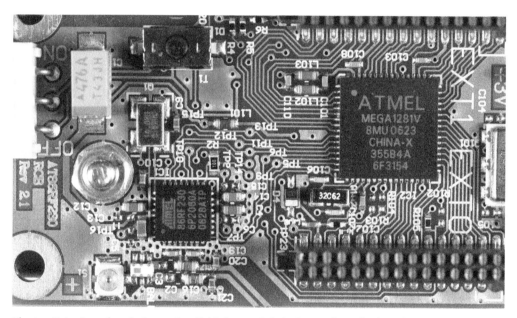

Photo 5.1: Just hook into the Z-Link module's 2mm female headers to get at the ATmega1281's free general-purpose I/O pins and access the AT86RF230 your way. The other side of the Z-Link module can be seen in Photo 5.2.

The Atmel AT86RF230

Low power, low cost, IEEE 802.15.4-compliant and ZigBee-ready are all words that describe the Atmel AT86RF230 IEEE 802.15.4-compliant transceiver. The idea behind the AT86RF230 is to provide the ZigBee or IEEE 802.15.4 network programmer/designer a nobrainer building-block path between a microcontroller's SPI pins and the IEEE 802.15.4 radio antenna hanging off of the AT86RF230.

In my opinion, RF design is black magic that is left to those that are in league with the beings of the underworld (you know, the Devil). In the case of the AT86RF230, all of the critical RF components with the exception of the antenna, crystal and decoupling capacitors are integrated into the AT86RF230 chip. So, you don't have to hold a degree from Devil U to put some IEEE 802.15.4 bits in the air with the AT86RF230.

The Z-Link AT86RF230 modules we will be using came as part of the Atmel ATAVR-RZ200 IEEE 802.15.4/ZigBee Demonstration Kit. In addition to five ATmega1281V-based AT86RF230 transceiver modules, the ATAVRRZ200 IEEE 802.15.4/ZigBee Demonstration Kit includes a Z-Link master module that is equipped with a very nice dot-matrix LCD and a socket for an AT86RF230-based Z-Link transceiver module. The ATAVRRZ200 IEEE 802.15.4/ZigBee Demonstration Kit Z-Link master module also provides hook-ups for programming and debugging the Z-Link master module's ATmega128 and the Z-Link module's ATmega1281V using an Atmel JTAGICE mkII or the ATAVRISP mkII programming dongle. As a convenience to the Z-Link developer (and if you just plain don't have one), the ATAVRRZ200 IEEE 802.15.4/ZigBee Demonstration Kit comes with an ATAVRISP mkII programming dongle. The ATAVRRZ200 IEEE 802.15.4/ZigBee Demonstration Kit Z-Link master module is lounging with a piggy-backed Z-Link module in Photo 5.2.

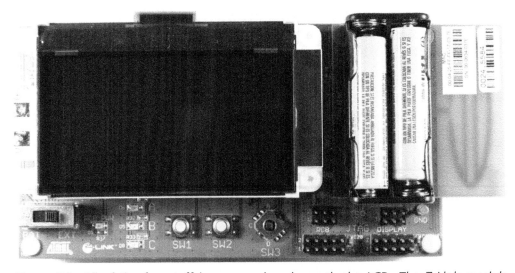

Photo 5.2: All of the fun stuff is crammed underneath the LCD. The Z-Link module piggybacks onto the Z-Link master controller board and can be seen at the extreme right of the shot. Everything is battery powered. However, the Z-Link master controller board has the capability of being mains powered.

A standard microcontroller SPI pin arrangement and four general-purpose I/O connections are all you need for access and control of the AT86RF230. The single-chip RF transceiver provides a complete radio interface between your favorite SPI-equipped microcontroller and the antenna connected to the AT86RF230's power amp.

As you can see in Figure 5.1, the AT86RF230 can be logically divided into an analog domain and a digital domain. The AT86RF230 transmits signals that are modulated using the O-QPSK method with half-sine pulse shaping and 32-length block spreading. O-QPSK (Offset-Quadrature Phase-Shift Keying) should ring a bell, as we discussed it in Chapter 3. Recall also that I recommended that you commit the entire set of 32-bit chip codes to memory. So, you shouldn't have to backtrack to refresh your memory with the O-QPSK concepts. Note the quadrature I and Q signals that are associated with O-QPSK in the AT86RF230's receive path in Figure 5.1. Since the I/Q pair appears in the receive chain of the AT86RF230, you can bet that the I/Q phase relationship method is also used in the AT86RF230's transmit chain.

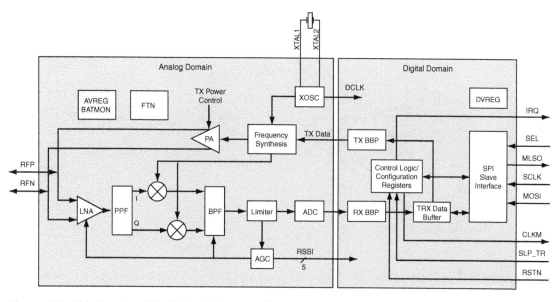

Figure 5.1: This is a simplified block diagram of the AT86RF230. Considering you've been exposed to it at least twice now, you should be able to hold your own in any I/Q modulation conversations that may pop up at the cocktail party.

AT86RF230 Modes of Operation

The AT86RF230 can operate in a number of modes that are suitable for supporting ZigBee operations. The AT86RF230 operating modes are controlled by only two signal pins and the SPI portal. When it comes to configuring an AT86RF230 operating mode two registers, TRX_STATE and TRX_STATUS, take the spotlight. The SPI feeds the TRX_STATE register and the success or failure of state or configuration changes is reflected in the TRX_STATUS register, which is also queried by the host microcontroller via the SPI portal.

The AT86RF230's active-high SLP_TR pin is used to enter SLEEP mode. It is also used to wake up the AT86RF230 transceiver. As you would imagine, SLEEP mode is the ultimate low-current consumption state of the AT86RF230. In fact, the only currents flowing during AT86RF230 SLEEP mode are leakage currents.

The AT86RF230's transceiver is forced into TRX_OFF mode by the AT86RF230's active-low RST pin unless the AT86RF230 is in the P_ON mode when RST is asserted. The relationship of P_ON to RST is depicted graphically in Figure 5.2.

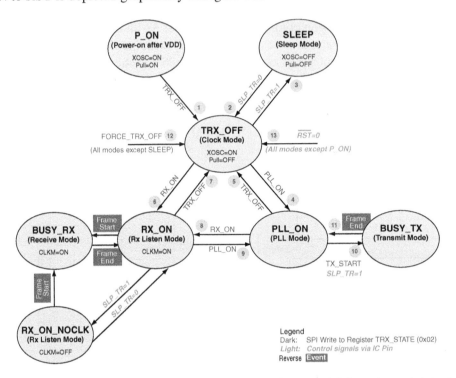

Figure 5.2: This figure seems very busy until you calm down and follow through it logically. All roads lead into and out of TRX_OFF.

Although the TRX_OFF and FORCE_TRX_OFF commands both take one down differing paths to an identical outcome, their resultant operation depends on the current state of the AT86RF230. If the AT86RF230's transceiver is in the BUSY_RX or BUSY_TX state, issuing FORCE_TRX_OFF will interrupt the receive or transmit process and force the AT86RF230 transceiver into the TRX_OFF state. Issuing a TRX_OFF command while the AT86RF230 is in a BUSY_TX or BUSY_RX state will not immediately kill the transmit or receive process that is in motion. Instead, the TRX_OFF command that was issued during the busy period is queued until the frame that is currently being transmitted or received is completely serviced. When the current transmit or receive operation is complete, the TRX_OFF command is executed. Another look at Figure 5.2 shows no direct TRX_OFF command paths between the BUSY_RX, BUSY_TX and TRX_OFF modes.

Let's walk through the rest of the basic AT86RF230 modes. As we fall through, bear in mind that any commands involving SLP_TR and RST are a result of a logic level applied to the associated AT86RF230 command input pin.

As you can plainly see in Figure 5.2, there is only one way to leave the P_ON mode. A TRX_OFF command must be issued by the host microcontroller via the SPI to transition from the initial power-on mode to the TRX_OFF state. When the AT86RF230 is powered up, the AT86RF230's crystal oscillator is activated and 128 μs later a master clock signal appears at the AT86RF230's CLKM pin. The small delay between crystal oscillator activation and the appearance of a clock signal at the AT86RF230's CLKM pin allows time for the AT86RF230's crystal oscillator to stabilize.

If the host microcontroller uses the CLKM signal as its system clock, the host microcontroller is said to be operating in synchronous mode with the AT86RF230. Pull-up resistors are connected to the AT86RF230's RST and SEL I/O pins and pull-down resistors are connected to the AT86RF230's SCLK, MOSI and SLP_TR pins. All of the pulling resistors are active in the P_ON state to keep logically undefined microcontroller general-purpose I/O pins from placing the AT86RF230 into an unwanted mode upon microcontroller power-up. The AT86RF230's pull-ups and pull-downs will be disabled when the AT86RF230's state is changed to TRX_OFF. Therefore, for proper operation of the AT86RF230 it is the duty of the microcontroller firmware to pull SLP_TR low and RST high before leaving the P_ON state.

Once the AT86RF230 is placed into the TRX_OFF state, the raising or lowering of the AT86RF230's SLP_TR signal pin will force the AT86RF230 to enter SLEEP mode or exit SLEEP mode, respectively. Entering AT86RF230 SLEEP mode totally disables the AT86RF230. Nothing electronic within the AT86RF230 functions during SLEEP mode, which explains why only leakage currents flow while the AT86RF230 sleeps. You can only put the AT86RF230 to sleep via the SLP_TR pin while in TRX_OFF mode and you can only wake up the AT86RF230 by lowering the SLP_TR pin to a logic 0 (zero) or resetting the AT86RF230 by bringing the AT86RF230's RST pin low. Applying a low to the AT86RF230's RST pin to exit sleep mode sets the AT86RF230's SPI and configuration registers to their default values and puts the AT86RF230 into the TRX_OFF state.

When the AT86RF230 is in the TRX_OFF state, the AT86RF230's SPI and the crystal oscillator are active. The digital domain of the AT86RF230 is also available as the AT86RF230's internal 1.8V voltage regulator is enabled. All pulling resistors are inactive while the AT86RF230 is in the TRX_OFF state.

The AT86RF230's analog voltage regulator is enabled when the AT86RF230 leaves the TRX_OFF state and enters the PLL_ON state. Upon activation, the PLL will eventually lock onto the receive frequency. Once this happens, an interrupt request will be signaled by the AT86RF230's IRQ pin. The host microcontroller can use this interrupt request signal as notification that the AT86RF230's PLL has successfully started and locked.

Entering the PLL_ON state from the TRX_OFF state is one of two paths that can be directly taken towards transmitting and receiving. An RX_ON command issued via the SPI pipe will move the AT86RF230 from the TRX_OFF state to the RX_ON state. Once the RX_ON state is reached, the AT86RF230's analog and digital receiver blocks and the PLL frequency synthesizer are energized. The AT86RF230 is in a listen mode at this point. When a preamble is detected, the digital receiver is activated and the AT86RF230's mode moves to BUSY_RX. If the microcontroller is using the CLKM signal for any reason, it will see an active CLKM signal in both the RX_ON and BUSY_RX modes.

The AT86RF230's basic modes make the general assumption that the AT86RF230's CLKM output is driving the host microcontroller in synchronous. That may or may not be the case depending upon the design. The idea behind having the host microcontroller use the CLKM signal is to leave the AT86RF230 on watch while the microcontroller sleeps. If the microcontroller is indeed being clocked by the AT86RF230, that's where the RX_ON_NOCLK state becomes very useful. RX_ON_NOCLK state is entered when SLP_TR is set to a logical high while the AT86RF230 is in the RX_ON mode. After SLP_TR goes high, the host microcontroller has 35 AT86RF230 clock cycles to do whatever it has to do to prepare for and go to sleep. Once the AT86RF230 goes into full-blown RX_ON_NOCLK mode and a frame creeps in while the microcontroller is snoozing, the AT86RF230 will produce an RX_START interrupt request. The AT86RF230 will then restart the clock signal at its CLKM pin and fall into the BUSY_RX state. When the dust settles and the incoming frame has been processed, the AT86RF230 will issue a TRX_END interrupt request via its IRQ pin and the AT86RF230's transceiver will enter the RX_ON state.

At this point, to reenter the RX_ON_NOCLK state from the RX_ON state, the SLP_TR pin must be reset to a logical 0 (zero) before being raised again to a logical 1, which will transition the AT86RF230 back into RX_ON_NOCLK state. As you have probably already deduced, resetting the AT86RF230's SLP_TR pin to 0 (zero) while in the RX_ON_NOCLK state will return the AT86RF230 to the RX_ON state.

From the TRX_OFF mode's point of view, the PLL_ON state is the jump-off point for initiating a transmission. The PLL_ON mode is also an indirect path that can be taken from the TRX_OFF mode to perform a receive operation as well. Once the PLL_ON state is entered, a transmission can be started by raising the SLP_TR pin to a logical 1 or sending a TX_START SPI command. No matter which way you go, the AT86RF230 will enter the BUSY_TX mode. The AT86RF230 does its thing during the BUSY_TX mode and when the frame has been transmitted, the AT86RF230 will return to the PLL_ON mode. If the SLP_TR pin was used to kick off the transmit operation, then the SLP_TR pin must be reset to 0 (zero) before being raised to a logical 1 to perform another transmit operation.

Stepping It Up a Notch

We've been talking about things like CSMA-CA and address filtering. Well, now is the time to put some faces on those words and acronyms I've been bantering about in the previous chapters.

The AT86RF230 supports two additional extended automatic modes, TX_ARET_ON and RX_AACK_ON, that augment the basic AT86RF230 modes we've just discussed. As you can see in Figure 5.3, the TX_ARET_ON and RX_AACK_ON modes are accessed via the basic RX_ON and PLL_ON states using SPI commands. The extended modes can also be accessed directly from the TRX_OFF basic mode using SPI commands. Return paths, which are also SPI commands, from the AT86RF230's automatic modes return the AT86RF230 to basic mode operation.

TX_ARET_ON is Atmel-ese for transmit/auto-retry. The TX_ARET_ON extended mode enables the transmission of a frame that has been, by way of the AT86RF230 MAC, automatically appended with a CRC. If the channel is busy and a transmission is pending or an expected ACK was not received following a transmission, the TX_ARET_ON mode will use the unslotted CSMA-CA algorithms and retry the transmit operation. A TX_ARET_ON interrupt request containing an exit code is issued upon the completion of a TX_ARET_ON transaction. The interrupt request exit code indicates the status of the transaction (success, no ACK, channel busy, etc.).

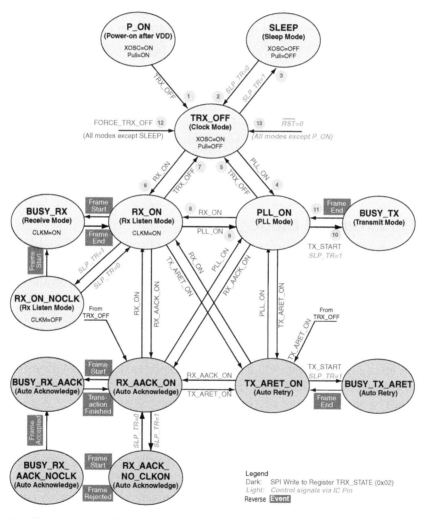

Figure 5.3: All we've done here is to add entry and exit pathways between the RX_ON and PLL_ON states to the AT86RF230's extended operating modes.

As the RX implies, the RX_AACK_ON (receive/auto-acknowledge) extended mode has to do with the reception of IEEE 802.15.4 frames. The RX_AACK_ON checks the CRC of incoming IEEE 802.15.4 frames and performs an address-filtering function. If the incoming frame has a good CRC, passes address filtering and an acknowledgment for the frame was requested, the RX_AACK_ON mechanisms will automatically transmit an ACK on behalf of the received frame.

An interrupt request is issued if a frame successfully gets by the AT86RF230's address filter. The automatic acknowledgment transmitted by the RX_AACK_ON mode is received by a peer node that should be operating in the TX_ARET_ON mode. The peer receiving the automatically generated ACK from the RX_AACK_ON node will check the incoming ACK for validity using the incoming ACK's CRC and sequence number. The sequence number of the incoming ACK frame should match the sequence number of the frame that requested the ACK.

Using the TX_ARET_ON and RX_AACK_ON extended modes is very much like using the AT86RF230's basic modes. In the extended modes, the TRX_STATE and TRX_STA-TUS registers are still the focal points of the SPI command structure. If you associate the RX_ON mode with the RX_AACK_ON mode, the RX_ON_NOCLK state with the RX_AACK_ON_NOCLK state, the PLL_ON mode with the TX_ARET_ON mode and the BUSY_TX state with the BUSY_TX_ARET state in Figure 5.3, you will notice that the results of using the SLP_TR pin logic levels in the basic mode and extended mode are identical. Just as we did with the basic mode, let's stroll through the automatic (extended operation) mode.

AT86RF230 Extended Mode

Recall that our very first pseudo-network using the ZMD44102 Starter Boards totally ignored the address fields of the IEEE 802.15.4 frames. That wasn't because the ZMD44102 transceiver was not capable of address filtering. It was a result of our turning off the ZMD44102's address-filtering capability in the application code. We actually loaded up all of the necessary addressing information into the ZMD44102 registers but we never allowed the ZMD44102 to use it. We did, however, enable and use the automatic ACK capability of the ZMD44102 transceiver. The point I'm trying to make here is that radios from differing manufacturers that are truly IEEE 802.15.4-compliant will have similar knobs to turn and switches to throw to allow them to interoperate on IEEE 802.15.4 and ZigBee networks. This is one of the reasons why ZigBee will be an even bigger success story as time passes.

It's rather obvious that the AT86RF230's address-filtering engine is enabled as RX_AACK_ON mode listens for incoming frames and parses incoming frames for frame type and destination address. An AT86RF230 operating in RX_AACK_ON mode will reject any frames with bad CRC bytes or an invalid destination address. An invalid destination address is like receiving mail that is not addressed to you. Hopefully, you give your incorrectly addressed mail back to the postman instead of trashing it like IEEE 802.15.4-compliant radios do. If a frame is accepted, a TRX_END interrupt request will be generated and the host microcontroller can then upload the frame. If the frame is capable of being acknowledged under IEEE 802.15.4 rules and the frame has requested an acknowledgment, the RX_AACK_ON engine will automatically send an ACK 12 symbol periods following the end of the received frame.

Transmission of IEEE 802.15.4 frames using the CSMA-CA algorithms that have been previously downloaded from the host microcontroller is the primary task of the TX_ARET_ON mode. Raising the SLP_TR pin high for a minimum of 1 µs starts a CSMA-CA transmission in TX_ARET_ON mode. The wording in the AT86RF230 datasheet does not mention the ability to kick off a CSMA-CA transmission in TX_ARET_ON mode using the SPI. However, the AT86RF230 state table in Figure 5.3 clearly indicates that a TX_START SPI command will also send the AT86RF230 into the BUSY_TX_ARET state from TX_ARET_ON mode. In any case, it is recommended that the data to be transmitted be downloaded from the host microcontroller before the CSMA-CA transmission process is started. You can also hold off and download the data to be transmitted while the AT86RF230 is clocking out the preamble bits if you have time to do so. If a clear channel is detected, the AT86RF230 will transmit the frame containing your preloaded data. If the surveillance of the transmitting channel detects traffic, the AT86RF230 transceiver will execute the CSMA-CA back-off algorithm and retry the transmission until the maximum number of retries has been exhausted. If the maximum retry count is reached and no successful transmission has occurred, the AT86RF230 transceiver will abort the transmission and indicate a transmission failure via a TRX_END interrupt request. The TRX_STATUS register's bits will be modified to convey the reason for the failure to the host microcontroller, which in this case would be CHANNEL_ACCESS_FAILURE.

As the AT86RF230 transceiver is transmitting the frame In TX_ARET_ON mode, it parses the outgoing frame to determine if an ACK reply is expected. If an ACK reply is expected, the transceiver switches into receive mode following the completion of the transmission and waits a specified time for the ACK reply frame. If no ACK reply is received within the ACK response time limitation, the entire CSMA-CA transmission process is repeated. Again, if the maximum retry limit is reached without a successful transmission/ACK reply sequence, the AT86RF230 aborts the transmission and posts a failure reason code of NO_ACK via a TRX_END interrupt request. If all goes as designed, a TRX_END interrupt request is generated and the TRX_STATUS register will be updated with a SUCCESS bit pattern.

The RX_AACK_NOCLK mode of operation is identical to the basic RX_ON_NOCLK mode with the exception of address filtering. In RX_AACK_NOCLK mode the AT86RF230's transceiver listens for the Start-of-Frame-Delimiter within an IEEE 802.15.4 frame. When a Start-of-Frame-Delimiter is detected, the AT86RF230 enters the BUSY_RX_AACK_NOCLK State and begins to receive the incoming frame. At this time the AT86RF230's CLKM pin is devoid of signal and if the host microcontroller is synchronizing with the AT86RF230 CLKM signal, it is most likely sleeping at this time. If the incoming frame passes the address-filter test, the AT86RF230 jumps into the BUSY_RX_AACK state, which enables the clock signal out of the AT86RF230's CLKM pin and allows a synchronized host microcontroller to process the incoming frame.

The AT86RF230 is a fascinating piece of IEEE 802.15.4 silicon. Although the operational concepts of the AT86RF230 are very easy to understand, to make sure that we have a good foundation to build upon, the folks at Atmel have provided us with a very nice set of AT86RF230 IEEE 802.15.4 MAC drivers. Let's use the things we've learned about the AT86RF230 and apply them to the Atmel IEEE 802.15.4 MAC firmware package.

Still, No Stack

I have five Atmel Z-Link modules, which are based on the AT86RF230. Instead of trying to run before we learn to walk, we are still only going to base our IEEE 802.15.4 network attempt on the pair of sublayers that are defined in the IEEE 802.15.4 standard, the PHY sublayer and the MAC sublayer. The Atmel IEEE 802.15.4 MAC firmware I alluded to earlier is called the Atmel IEEE 802.15.4 MAC. The functionality of the Atmel IEEE 802.15.4 MAC firmware is contained within a library called libl2_rdk230. The Atmel IEEE 802.15.4 MAC library follows the IEEE 802.15.4 standard and is aimed at providing services for the ZigBee NWK layer, which lies just above the MAC sublayer in a ZigBee stack.

The Atmel IEEE 802.15.4 MAC can be utilized in both FFD and RFD nodes. To make that happen, all you need are standard AVR programming/debugging tools to put the Atmel IEEE 802.15.4 MAC to work. I happen to have a set of those standard AVR tools and I'm all about putting the Atmel IEEE 802.15.4 MAC to work with the five Z-Link nodes in my possession.

It should be rather obvious that the Atmel IEEE 802.15.4 MAC is designed to support the Atmel AT86RF230 IEEE 802.15.4-compliant transceiver IC. The Atmel IEEE 802.15.4 MAC BIOS/HAL library code sets up the ATmega1281 microcontroller's general-purpose I/O and SPI to directly interface to the AT86RF230's I/O configuration.

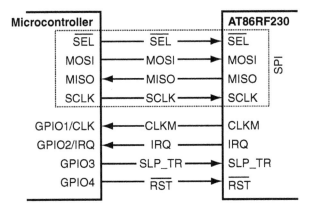

Figure 5.4: The AT86RF230 operates as an SPI slave. Two additional general-purpose I/O signals are used in conjunction with the SPI signals by the AT86RF230 to determine what mode the AT86RF230 will operate in.

The Atmel IEEE 802.15.4 MAC code communicates with the AT86RF230's internals via the ATmega1281's SPI engine as shown in Figure 5.4. AT86RF230 register programming and frame-transfer operations are performed using the SPI interface. The AT86RF230 is equipped with an internal 128-byte frame buffer, which allows the AT86RF230 to buffer one transmit or one receive frame in a mutually exclusive manner. To those of you reading this book that are from my part of the Southern United States, that mutually exclusive thing just means one at a time.

In addition to the SPI signals, the AT86RF230 requires four additional host microcontroller general-purpose I/O lines to enable the full capability of the AT86RF230 IEEE 802.15.4-compliant transceiver circuitry. As you already know, the AT86RF230 CLKM pin supplies a clock signal that is generated by the AT86RF230. The CLKM signal can be used by the host microcontroller as a system clock in synchronous mode or as a timing reference in asynchronous mode.

The Atmel Z-Link radio modules are built around a combination of the AT86RF230 IEEE 802.15.4-compliant transceiver IC and the ATmega1281 microcontroller. The ATmega1281 can easily contain an entire IEEE 802.15.4 or ZigBee application within its 128 KB of self-programmable program Flash memory. As IEEE 802.15.4 and ZigBee packet sizes max out at 127 bytes, there is also more than enough room for IEEE 802.15.4 packet buffers and any other of the application's volatile storage needs inside the ATmega1281's 8 KB of SRAM space. ZigBee and IEEE 802.15.4 things such as 64-bit IEEE addresses and network-configuration information sometimes need to be stored in nonvolatile storage areas. The ATmega1281 is a good choice for use in IEEE 802.15.4 and ZigBee networks as it has 4 KB of nonvolatile EEPROM storage space. You can also burn the 64-bit IEEE 802.15.4 address into the ATmega1281's program Flash if your application calls for that.

The functionality provided by the Atmel IEEE 802.15.4 MAC firmware is contained within a library file that we can access using API calls that are named after the primitives they represent. Figure 5.5 is a conceptual look at the Atmel IEEE 802.15.4 MAC library.

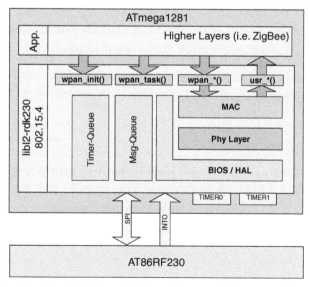

Figure 5.5: BIOS is short for Basic Input-Output System. One normally associates the term BIOS with personal computers. However, a microcontroller can also wield a BIOS. HAL is not the computer that you saw (and listened to a lot) in the movie 2001: A Space Odyssey. *HAL means Hardware Abstraction Layer in the AT86RF230 universe. In the case of the AT86RF230, the BIOS/HAL combination is the code that integrates the AT86RF230 and the ATmega1281 at the hardware level.*

Just in case you're wondering why a PHY Layer box is embedded within a MAC library block diagram, the Atmel IEEE 802.15.4 MAC contains PLME calls that perform PHY functions like CCA (Clear Channel Assessment), ED (Energy Detection) and PHY PIB attribute PLME SET and PLME GET requests.

The Atmel IEEE 802.15.4 MAC uses its internal Timer-Queue to give the AT86RF230 programmer access to timers that can be started and stopped on command. When a timer expires, a callback function is invoked to notify the programmer that the timer has expired. It is the responsibility of the programmer to service the timer callback function.

The message queue (Msg-Queue) lines up messages that are sent to the MAC sublayer. Messages arriving in the MAC layer can generate new messages, which get added to the message queue, or kick off local operations. The wpan_task function is responsible for processing messages in the message queue. Therefore, the wpan_task function must be called as often as possible. Whenever the wpan_task function processes a message in the message queue, it will return a Boolean TRUE. Otherwise, the wpan_task function will return a Boolean FALSE. All wpan_-prefixed functions are generated outside of the MAC sublayer and usually come from the application layer or another layer on top of the MAC sublayer.

Since most of the wpan_-prefixed function calls are associated with the methods of passing a primitive, depending upon the service represented by the primitive, some type of primitive-like confirmation response may be required. In many cases, when a request primitive is issued, a confirmation primitive may be issued in response. If I say that in Atmel IEEE 802.15.4 MAC-ese, when a wpan_-prefixed message is placed in the message queue that invokes a return message, the return message will be in the form of a callback function prefixed by user_.

While we're on the subject of functions, the Atmel IEEE 802.15.4 MAC library supports three types of functions:

- General Functions
- Request Functions
- Callback Functions

General Functions are used to initialize and process the Atmel IEEE 802.15.4 MAC. Access to the Atmel IEEE 802.15.4 MAC library's timer entity is also a task handled by the General Functions. Atmel IEEE 802.15.4 MAC General Functions include:

- wpan_init
- wpan_task
- wpan_start_timer
- wpan_stop_timer
- usr_timer_trigger

Atmel IEEE 802.15.4 MAC initialization is performed by the wpan_init function. This function is called (and must be called) before any other function and initializes all of the library resources.

The Atmel IEEE 802.15.4 MAC firmware requires that we constantly call the wpan_task function as fast as we can, as the wpan_task function is in control of dispatching and servicing the message traffic that is flowing into the library's message queue. The wpan_task function returns a Boolean TRUE when a message is processed.

The library timer entity is brought into the fray when the wpan_start_timer function is set into motion. Upon invocation of the wpan_start_timer function, any of the individual timers identified by a number between 0 and 255 are started for a specified number of symbol periods as called out in the wpan_start_timer function arguments. To stop any timer that has been started with the wpan_start_timer function, a call to the wpan_stop_timer function must be issued.

The usr_timer_trigger function is a callback function that is called for each user-defined timer that expires. The application stuffs this function with the necessary code needed to respond to the timer expiration event. The individual timers are identified with a number in the range of 0–255.

All of the Atmel IEEE 802.15.4 MAC Request Functions, with the exception of a couple, are MLME-oriented. Atmel IEEE 802.15.4 MAC Request Functions enable the programmer to send messages within the Atmel IEEE 802.15.4 MAC.

There are two MCPS-related functions supported by the Atmel IEEE 802.15.4 MAC. Recall that MCPS (MAC Common Part Sublayer) primitives represent data-handling services within the MAC sublayer. The Atmel IEEE 802.15.4 MAC's

wpan_mcps_data_request function forms an MCPS Data Request message and inserts it into the Atmel IEEE 802.15.4 MAC message queue. The other wpan_mcps_-prefixed function, wpan_mcps_purge_request, creates an MCPS Purge Request message that gets dropped into the Atmel IEEE 802.15.4 MAC message queue for round-robin servicing by the wpan_task function.

These MLME messages, which are all prefixed by wpan_mlme_, may be used to control connection establishment, scanning of the available channels, association and disassociation of network nodes and disconnections from the network. There are also Atmel IEEE 802.15.4 MAC wpan_mlme_-prefixed functions that access and alter the MAC PIB database attributes. There are many MLME functions supported by the Atmel IEEE 802.15.4 MAC. Rather than try to rattle them all off here, we'll deal with them as we encounter them in the working network code we're about to explore in Code Snippet 5.1.

An AT86RF230 PAN Coordinator Application

A PAN Coordinator application lies within the lines of code you see in Code Snippet 5.1. There is a complementary End Device application that we will examine when we're finished absorbing the intent of the PAN Coordinator application.

Code Snippet 5.1

```
*************************************************************************
/*====Coordinator Application============================== */
#  define _BV(x) (1<<(x))
#include "wpan_defines.h"
#include "ieee_const.h"
#include "ieee_types.h"
#include "wpan.h"

/*====Macros=================================== */
#ifndef RF_CHANNEL
# define RF_CHANNEL (16)
# warning "RF channel undefined, using channel 16 as default"
#endif

#define PANCOORD_SHORT_ADD   (0xBABE)
#define BROADCAST_SHORT_ADD (0xFFFF)
#define PANID               (0xCAFE)

#define PANCOORD_SCAN_CHANNELS  ((uint32_t)0x00000000)
#define SCANDURATION                  (3)
#define BEACON_ORDER                  (15)    /*  NO Beacon */
#define SUPERFRAME_ORDER              (15)
#define EMPTY_LONG_ADDRESS  (~(uint64_t)0)

#define MAX_ENTRIES      (8)
#define NO_ENTRY         (0xFF)

/* macro stores the state value and sets the state led to 0 */
#define SET_STATE(x) do { c_status.state=(x); \
                          PORTE |= (_BV((uint8_t)(x)));} while(0)

/*===Typedefs=================================== */
typedef struct
{
    bool associated;
    uint64_t long_addr;
} association_entry_t;

typedef enum
{
    INIT_DONE,
    PEND_RESET,
    PEND_SET_SHORT_ADDR,
    PEND_INITIAL_SCAN,
    PEND_START,
    PEND_ASSOC_PERMIT,
    RUN,
} coord_state_t;

typedef struct
```

```
{
    uint8_t handle;
    uint8_t led_value;
    coord_state_t state;
} coord_status_t;


/*===StaticVariables==================================== */
static coord_status_t c_status;
static association_entry_t association_table[MAX_ENTRIES];

uint8_t data_buffer[127];

/* === Prototypes ===================================== */
void application_init(void);
void mac_do_reset(void);
void mac_active_scan(void);
void mac_start_pan(void);
void mac_set_short_addr(uint16_t addr);
void mac_set_assoc_permit(uint8_t permit);
void mac_register_device(uint64_t DeviceAddress);
void mac_send_data(void);

static void association_table_init(void);
static uint8_t get_empty_association_slot(void);
static uint8_t search_association_entry(uint64_t addr);

/* === Implementation ================================== */
int main(void)
{
    application_init();

    mac_do_reset();

    while(1)
    {
        while(wpan_task())
        {
        }
    }
}


void application_init(void)
{

    c_status.led_value = 0;
    c_status.handle = 0;
    association_table_init();
```

```
    /* init IO ports */
    DDRE = 0xDF;  /* PE5 is input */
    PORTE = 0x00; /* all LED's ON */

    /* init mac layer */
    wpan_init();
    SET_STATE(INIT_DONE);

    /* enable interrupts */
    sei();
    return;
}

void mac_do_reset()
{
    wpan_mlme_reset_request( true );
    SET_STATE( PEND_RESET );
}


void mac_set_short_addr(uint16_t addr)
{
    wpan_mlme_set_request (macShortAddress, &addr, sizeof(addr));
    SET_STATE ( PEND_SET_SHORT_ADDR );
}


void mac_set_assoc_permit(uint8_t permit)
{
    wpan_mlme_set_request (macAssociationPermit, &permit, sizeof(permit));
    SET_STATE ( PEND_ASSOC_PERMIT );
}


void usr_mlme_reset_conf (uint8_t status)
{
    if ((status == MAC_SUCCESS) && (c_status.state == PEND_RESET))
    {
        mac_set_short_addr(PANCOORD_SHORT_ADD);
    }
    else
    {
        mac_do_reset();
    }
    return;
}

void usr_mlme_set_conf(uint8_t status, uint8_t PIBAttribute )
{
    switch (c_status.state)
```

```
    {
        case PEND_SET_SHORT_ADDR:
            if ((status == MAC_SUCCESS) && (PIBAttribute ==
macShortAddress))
                {
                    mac_active_scan();
                }
            break;
        case PEND_ASSOC_PERMIT:
            if ((status == MAC_SUCCESS) && (PIBAttribute ==
macAssociationPermit))
                {
                    SET_STATE( RUN );
                    PORTE = 0xff; /* LED's off if we come to here */
                }
        default:
            break;
    }

    return;
}


void mac_active_scan(void)
{
    wpan_mlme_scan_request (MLME_SCAN_TYPE_ACTIVE, PANCOORD_SCAN_CHANNELS,
SCANDURATION);
    SET_STATE(PEND_INITIAL_SCAN);
    return;
}


void usr_mlme_scan_conf(uint8_t status, uint8_t ScanType, uint32_t
UnscannedChannels,
                        uint8_t ResultListSize, uint8_t *data, uint8_t data_
length)
{
    if (c_status.state == PEND_INITIAL_SCAN)
    {
        /* We don't care about the confirm of the scan request because the
scan */
        /* request just puts the MAC into the correct state for a pancoord.
*/
        mac_start_pan();
    }
    return;
}


void mac_start_pan(void)
```

```
{
    wpan_mlme_start_request ( PANID, RF_CHANNEL,
                               BEACON_ORDER, SUPERFRAME_ORDER,
                               true, false, false, false);
    SET_STATE(PEND_START);
    return;
}


void usr_mlme_start_conf(uint8_t status)
{
    if  (c_status.state == PEND_START)
    {
        if (status == MAC_SUCCESS)
        {
            mac_set_assoc_permit( 1 );
        }
        else
        {
            mac_start_pan();
        }
    }
    return;
}


void association_table_init(void)
{
    for(uint16_t i = 0; i < MAX_ENTRIES; i++)
    {
        association_table[i].associated = false;
        association_table[i].long_addr = EMPTY_LONG_ADDRESS;
    }
    return;
}


uint8_t get_empty_association_slot(void)
{
    uint8_t ret = NO_ENTRY, idx;

    for(idx = 0; idx < MAX_ENTRIES; idx++)
    {
        if(association_table[idx].long_addr == EMPTY_LONG_ADDRESS)
        {
            ret = idx;
            break;
        }
    }

    return ret;
}
```

```
uint8_t search_association_entry(uint64_t addr)
{
uint8_t ret = NO_ENTRY, idx;
    /*  Look for the address in the table. */
    for(idx = 0; idx < MAX_ENTRIES; idx++)
    {
        if(association_table[idx].long_addr == addr)
        {
            ret = idx;
            break;
        }
    }

    return ret;
}


void usr_mlme_associate_ind (uint64_t DeviceAddress, uint8_t
CapabilityInformation,
                            uint8_t SecurityUse, uint8_t ACLEntry)
{
    if (c_status.state == RUN)
    {
        mac_register_device(DeviceAddress);
    }
}


void mac_register_device(uint64_t DeviceAddress)
{
    uint8_t entry_index;

    /*  Search if device is already associated. */

    entry_index = search_association_entry(DeviceAddress);
    if(entry_index == NO_ENTRY)
    {
        entry_index = get_empty_association_slot();
    }

    if (entry_index == NO_ENTRY)
    {
        wpan_mlme_associate_response (DeviceAddress, 0,
            PAN_AT_CAPACITY, false);
    }
    else
    {
        association_table[entry_index].long_addr = DeviceAddress;
        wpan_mlme_associate_response(DeviceAddress, entry_index,
            ASSOCIATION_SUCCESSFUL, false);
    }
    return;
```

```
}


void usr_mlme_comm_status_ind(wpan_commstatus_addr_t *pAddrInfo, uint8_t
status)
{
uint16_t assoc_idx = NO_ENTRY;

    if (pAddrInfo->PANId == PANID) /* if it is our PAN */
    {
        if (pAddrInfo->DstAddrMode == WPAN_ADDRMODE_SHORT)
        {
            assoc_idx = pAddrInfo->DstAddr;
        }
        else if (pAddrInfo->DstAddrMode == WPAN_ADDRMODE_LONG)
        {
            assoc_idx = search_association_entry(pAddrInfo->DstAddr);
        }

        if ((status == MAC_SUCCESS) && (assoc_idx < MAX_ENTRIES))
        {
            association_table[assoc_idx].associated = true;
        }
    }
}


void usr_mcps_data_ind (wpan_mcpsdata_addr_t *addrInfo, uint8_t
mpduLinkQuality,
                        uint8_t SecurityUse, uint8_t ACLEntry,
                        uint8_t msduLength, uint8_t *msdu)
{
    /* Determine if the indication comes from a device,
     * that has previously associated.
     */
    if (addrInfo->SrcAddrMode == WPAN_ADDRMODE_SHORT
        && association_table[(uint8_t)addrInfo->SrcAddr].associated)
    {
        /*  Mask the data byte with the address info. */
        if(msdu[0])
        {
            c_status.led_value |= (1 << (uint8_t)addrInfo->SrcAddr);
        }
        else
        {
            c_status.led_value &= ~(1 << (uint8_t)addrInfo->SrcAddr);
        }

        /*  Show the address of the device on our LEDs. */
        PORTE = ~c_status.led_value;

        /*  Start sending data to all associated devices. */
```

```
            mac_send_data();
    }
    return;
}


void mac_send_data(void)
{
    wpan_mcpsdata_addr_t addr_info;
    addr_info.SrcAddrMode = WPAN_ADDRMODE_SHORT;
    addr_info.SrcPANId = PANID;
    addr_info.SrcAddr = PANCOORD_SHORT_ADD;
    addr_info.DstAddrMode = WPAN_ADDRMODE_SHORT;
    addr_info.DstPANId = PANID;
    addr_info.DstAddr = BROADCAST_SHORT_ADD;
    wpan_mcps_data_request( &addr_info, c_status.handle++, WPAN_TXOPT_OFF,
                            (void *)&c_status.led_value,
                            sizeof(c_status.led_value));
    return;
}
```
****

Code Snippet 5.1: This code comprises a full-blown IEEE 802.15.4 application. Following through the flow of this code should drive home the principles behind primitives.

The PAN Coordinator application basically establishes and starts an IEEE 802.15.4 network. Once the network is up and running, End Devices that desire to can join (associate with) the newly founded IEEE 802.15.4 network. Each end device is running an end device application that simply looks at the state of a pushbutton switch. The associated end devices will send a frame to the PAN Coordinator every time a switch closure is detected. The PAN Coordinator will in turn receive the switch data and disperse it to all of the associated end device nodes. Let's tear down the code module by module and line by line.

Once you get the idea behind what the Atmel IEEE 802.15.4 MAC is actually doing, the code is a bunch of fun to flow through, as essential IEEE 802.15.4 network concepts are revealed with simple coding algorithms. The initial #include statements you see in Code Snippet 5.2 pull in all of the necessary IEEE 802.15.4 constant and type definitions that correspond to the constants you find in the IEEE 802.15.4 standard document.

Studying the contents of the #include files will shed lots of light on IEEE 802.15.4 intentions and the role of primitives. The really neat thing about the ieee_const.h file is that the location of the corresponding IEEE 802.15.4 definition within the IEEE 802.15.4 standard document is often bundled in with the associated C definition statement. An example of this is shown in Code Snippet 5.2.

Code Snippet 5.2
****
```
/*====Coordinator Application============================= */
#include "wpan_defines.h"
#include "ieee_const.h"
```

```
#include "ieee_types.h"
#include "wpan.h"

/*====From ieee_const.h =============================== */
#define aMaxPHYPacketSize        (127) // maximum size of PHY packet
#define aMaxFrameOverhead   (25)

/* 7.4.1 MAC Layer Constants */
/**
 * The maximum number of octets that can be
 * transmitted in the MAC frame payload field.
 *
 * @ingroup apiMacConst
 */
#define aMaxMACFrameSize              (aMaxPHYPacketSize -
aMaxFrameOverhead)
```
****

Code Snippet 5.2: If you browse the IEEE 802.15.4 standard document and search for 7.4.1, you'll find yourself in section 7.4 MAC constants and PIB attributes. Subsection 7.4.1 points you to a table of MAC constants.

You'll find primitive definitions and their callback function definitions and descriptions in the wpan.h include file. Again, to make it a bit easier to get on top of things, IEEE 802.15.4 standard document locations are supplied within the wpan.h function descriptions to augment the information that is given in the text about the function. I've pulled out a wpan_mlme_set_request function and its associated callback function usr_mlme_set_conf as an example of this in Code Snippet 5.3.

Code Snippet 5.3
/
```
 * Forms an MLME SET REQUEST message and puts it in the message queue.
 * @param PIBAttribute The PIB attribute to be set (see @ref apiMacMib).
 * @param PIBAttributeValue A void pointer which points to the value to be
stored in the PIB attribute.
 * @param PIBAttributeValueSize The size of PIBAttributeValue.
 * @return true = success, false = failed to add to message queue because of
overflow.
 */
bool wpan_mlme_set_request(const uint8_t PIBAttribute,
    const void *PIBAttributeValue, const size_t PIBAttributeValueSize);

/**
 * @brief Callback function for an mlme_set_conf message.
 *
 * This function has to be implemented by the application in order to
 * process a message of the type mlme_set_conf_t coming from the stack.
 *
 * @param status
 *     The result code for the corresponding request (see 7.1.14.2.1 in
802.15.4-2003).
```

```
 * @param PIBAttribute
 *     The identifier of the PHY PIB attribute to get (see @ref apiMacMib).
 *
 * @return void
 *
 * @ingroup apiMacCb
 */
void usr_mlme_set_conf(uint8_t status, uint8_t PIBAttribute);
```
****

Code Snippet 5.3: A rare typo in the Atmel listing, 7.1.14.2.1 should be 7.1.13.2.1 in this snippet. Putting the 13 in there takes you to the right place, which is the service primitive semantics subsection of the MLME-SET.confirm section of the IEEE 802.15.4 standard document. Regardless of the typo, I think you get the idea.

The PAN definitions in the Macros area of the Atmel IEEE 802.15.4 MAC are very creative. After seeing the PANID and PAN short address in Code Snippet 5.4, I found myself trying to assemble as many words as I could using only the hex letters A–F.

Code Snippet 5.4
****
```
/*===Macros========================================= */
#ifndef RF_CHANNEL
# define RF_CHANNEL (16)
# warning "RF channel undefined, using channel 16 as default"
#endif

#define PANCOORD_SHORT_ADD      (0xBABE)
#define BROADCAST_SHORT_ADD     (0xFFFF)
#define PANID                   (0xCAFE)

#define PANCOORD_SCAN_CHANNELS  ((uint32_t)0x00000000)
#define SCANDURATION                (3)
#define BEACON_ORDER                (15)    /*  NO Beacon */
#define SUPERFRAME_ORDER            (15)
#define EMPTY_LONG_ADDRESS  (~(uint64_t)0)

#define MAX_ENTRIES      (8)
#define NO_ENTRY         (0xFF)

/* macro stores the state value and sets the state led to 0 */
#define SET_STATE(x) do { c_status.state=(x); \
                    PORTE |= (_BV((uint8_t)(x)));} while(0)
```
****

Code Snippet 5.4: You can't get tricky with the broadcast short address as 0xFFFF is the only option.

IEEE 802.15.4 short addresses are 16 bits in length. An official IEEE 802.15.4 address is 64 bits in length. By sending only 16 address characters, short addresses reduce transmission time, thus saving power. Using short addresses also allows more payload data to be packed into a 127-byte frame.

Note that the Beacon order and superframe order definitions in Code Snippet 5.4 are set for decimal 15. That disables the emission of periodic Beacons and thus eliminates superframe generation, which means an unslotted network will be formed by this PAN Coordinator. We could easily create a slotted network by reducing the superframe order and Beacon order values below decimal 15.

The Typedefs area coded in Code Snippet 5.5 provides us with the data structures that ultimately provide the SRAM holding areas for the association table and status variables. Note the states that are enumerated in the coord_state_t typedef. These states are important as they help guide the flow of the Atmel MAC firmware's task execution. We'll be able to see into the variable values when we run the PAN Coordinator code in debug mode.

Code Snippet 5.5

```
**************************************************************************
/*===Typedefs========================================= */
typedef struct
{
    bool associated;
    uint64_t long_addr;
} association_entry_t;

typedef enum
{
    INIT_DONE,
    PEND_RESET,
    PEND_SET_SHORT_ADDR,
    PEND_INITIAL_SCAN,
    PEND_START,
    PEND_ASSOC_PERMIT,
    RUN,
} coord_state_t;

typedef struct
{
    uint8_t handle;
    uint8_t led_value;
    coord_state_t state;
} coord_status_t;
**************************************************************************
```

Code Snippet 5.5: The enumerated (numbered) states are used to control the flow of the PAN Coordinator application. If you're not a C coder, this may be intimidating. Once you see what the variables hold in a debug window, the intimidation factor will fall to zero.

The actual SRAM tables are simple data structures that are built using the templates presented in the PAN Coordinator's Typedefs code. The code in Code Snippet 5.6 defines a status data structure (c_status) that will use the enumerated states of coord_state_t to provide us with the current application run state among other things.

Code Snippet 5.6

```
*********************************************************************
/*===StaticVariables================================== */
static coord_status_t c_status;
static association_entry_t association_table[MAX_ENTRIES];

uint8_t data_buffer[127];
*********************************************************************
```

Code Snippet 5.6: These are simply SRAM definitions that will setup areas of SRAM according to the data structure templates in the Typedefs code. Note the length of the data buffer. Does 127 bytes ring a bell?

According to Code Snippet 5.6, an association table array containing eight entries will be allocated in the ATmega1281's SRAM. Each association table entry will include a TRUE/FALSE indication of association and the official IEEE 802.15.4 64-bit address of the associated node.

Recall that the functionality provided by the Atmel IEEE 802.15.4 MAC firmware is contained within a library file that we can access using API calls that are named after the primitives they represent. Any API call beginning with wpan_ is aimed at the Atmel IEEE 802.15.4 MAC library and if any action generated by the API call needs to be handled by the Atmel IEEE 802.15.4 MAC firmware, it will be given back to us in a callback function that begins with user_.

The main function of the PAN Coordinator is coded up in Code Snippet 5.7. Following the initialization of the application variables and the MAC reset, the PAN Coordinator application spins through the wpan_task function as fast as it can while servicing other tasks in a round-robin fashion.

Code Snippet 5.7

```
*********************************************************************
/* === Implementation ================================== */
int main(void)
{
    application_init();

    mac_do_reset();

    while(1)
    {
        while(wpan_task())
        {
//low execution time tasks call from here
        }
//longer execution time tasks call from here
    }
}
*********************************************************************
```

Code Snippet 5.7: We want the wpan_task function to spin as much as possible. So, Atmel recommends that tasks that can be completed quickly be called from within the wpan_task function braces. Tasks that require a bit more processing time should be called outside of the wpan_task function's braces.

Before we can turn on the wpan_task spin cycle, we need to make sure all of our ducks are in a row on the hardware and SRAM fronts. That's why we call the application_init function right off the bat.

The very first thing we do in Code Snippet 5.8 is to initialize the c_status data structure. As you can see, we basically set everything within the c_status data structure to 0 (zero) using the numeric value of 0 (zero), the Boolean FALSE and EMPTY_LONG_ADDRESS definitions.

Code Snippet 5.8

```
/* ================================================ */
void application_init(void)
{

    c_status.led_value = 0;
    c_status.handle = 0;
    association_table_init();

    /* init IO ports */
    DDRE = 0xFF;  /* all bits of PORT E are outputs */
    PORTE = 0x00; /* all LED's ON */

    /* init mac layer */
    wpan_init();
    SET_STATE(INIT_DONE);

    /* enable interrupts */
    sei();
    return;
}

void association_table_init(void)
{
    for(uint16_t i = 0; i < MAX_ENTRIES; i++)
    {
        association_table[i].associated = false;
        association_table[i].long_addr = EMPTY_LONG_ADDRESS;
    }
    return;
}
/* ================================================ */
```

Code Snippet 5.8: This code clears the ATmega1281 SRAM holding areas, initializes the MAC and extinguishes the LEDs. We want to have all of the association tables and run states clean before we start feeding the Atmel IEEE 802.15.4 MAC's message queue.

According to Schematic 5.1, the Z-Link's ATmega1281 is hosting a trio of LEDs on its general-purpose I/O pins PE2, PE3 and PE4 and a pushbutton switch on PE5. All of the LEDs are turned on in the initialization routine.

From an earlier discussion we know that we must call the wpan_init function before we call any other Atmel IEEE 802.15.4 MAC library function. After calling wpan_init, we can now set the PAN Coordinator application state to INIT_DONE and enable the ATmega1281's interrupts.

It may be a good idea to make sure that the AT86RF230's MAC is ready to work as designed before we start using it. So, let's follow the execution of the mac_do_reset function in Code Snippet 5.9.

Code Snippet 5.9

```
**************************************************************************
void mac_do_reset()
{
    wpan_mlme_reset_request( true );
    SET_STATE( PEND_RESET );
}
**************************************************************************
```

Code Snippet 5.9: This is the Atmel IEEE 802.15.4 MAC's way of issuing a MLME-RESET.request primitive. The PAN Coordinator application state is changed so that the callback function can use the PAN Coordinator application's state in its decision as to the status of the execution of the wpan_mlme_reset_request function.

The first thing the mac_do_reset function does is place an API call to wpan_mlme_reset_request with the function argument set to Boolean TRUE. The mlme in the call stands for MAC Layer Management Entity, which is the management part of the MAC layer that is accessed via the management SAP (MLME-SAP). We have effectively used MLME-SAP to pass a primitive to the MAC's management area (MLME) asking for the invocation of a service to reset the MAC.

So that the PAN Coordinator application knows what was done prior, the PAN Coordinator application's state is changed from INIT_DONE to PEND_RESET in the application. Our callback function for the wpan_mlme_reset_request function is usr_mlme_reset_conf (MLME-RESET.confirm in the IEEE 802.15.4 primitive form) and can be viewed in Code Snippet 5.10. The conf is short for confirm, which is another common word from the IEEE 802.15.4 primitive language we've been speaking since you opened this book.

Code Snippet 5.10

```
**************************************************************************
void usr_mlme_reset_conf (uint8_t status)
{
    if ((status == MAC_SUCCESS) && (c_status.state == PEND_RESET))
    {
        mac_set_short_addr(PANCOORD_SHORT_ADD);
    }
    else
    {
        mac_do_reset();
    }
```

```
    return;
}

void mac_set_short_addr(uint16_t addr)
{
    wpan_mlme_set_request (macShortAddress, &addr, sizeof(addr));
    SET_STATE ( PEND_SET_SHORT_ADDR );
}
```
****

Code Snippet 5.10: The PEND_RESET state tells the callback function that the wpan_mlme_reset_request function was previously visited. The mac_do_reset function is the only place the wpan_mlme_reset_request function is called and the only place PEND_RESET is set as a state. This is beginning to be fun, isn't it? Why is it fun? Because it all interlocks and makes sense.

I never assume, as it turns you into a donkey. So, let's pretend that things went as planned within the wpan_mlme_reset_request function and the status returned by our wpan_mlme_reset_request API call was MAC_SUCCESS. We have previously set the new state to PEND_RESET, which satisfies the remainder of the "if" statement in the usr_mlme_reset_conf callback function and invokes the mac_set_short_addr function, which calls the wpan_mlme_set_request function to set the PAN Coordinator's short address to 0xBABE. If we pretend to let things go sour here, the MAC reset is attempted again.

Back at the ranch, the PAN Coordinator application's state gets changed to PEND_SET_SHORT_ADDR within the mac_set_short_addr function. Got the idea? A request API call will most always be answered by a user confirm callback function. IEEE 802.15.4 primitives work in the same way.

The wpan_mlme_set_request function triggers the callback function usr_mlme_set_conf in Code Snippet 5.11.

Code Snippet 5.11
****
```
void usr_mlme_set_conf(uint8_t status, uint8_t PIBAttribute )
{
    switch (c_status.state)
    {
        case PEND_SET_SHORT_ADDR:
            if ((status == MAC_SUCCESS) && (PIBAttribute == macShortAddress))
            {
                mac_active_scan();
            }
            break;
        case PEND_ASSOC_PERMIT:
            if ((status == MAC_SUCCESS) && (PIBAttribute == macAssociationPermit))
            {
                SET_STATE( RUN );
                PORTE = 0xff; /* LED's off if we come to here */
            }
        default:
```

```
                break;
        }

        return;
}
```
****

Code Snippet 5.11: This piece of code shows you why the states within the PAN Coordinator application are so important. The PAN Coordinator application will perform a channel scan or go into RUN state depending on the PAN Coordinator application's state when this callback is executed.

Since our PAN Coordinator application state is currently set to PEND_SET_SHORT_ADDR, the mac_active_scan function will be called in Code Snippet 5.11. The wpan_mlme_scan_request function will scan looking for inactive channels.

Code Snippet 5.12
****

```
void mac_active_scan(void)
{
    wpan_mlme_scan_request (MLME_SCAN_TYPE_ACTIVE, PANCOORD_SCAN_CHANNELS,
SCANDURATION);
    SET_STATE(PEND_INITIAL_SCAN);
    return;
}


void usr_mlme_scan_conf(uint8_t status, uint8_t ScanType,
uint32_t UnscannedChannels,
uint8_t ResultListSize, uint8_t *data,
uint8_t data_length)
{
    if (c_status.state == PEND_INITIAL_SCAN)
    {
        /* We don't care about the confirm of the scan request because the scan */
        /* request just puts the MAC into the correct state for a pancoord. */
        mac_start_pan();
    }
    return;
}


void mac_start_pan(void)
{
    wpan_mlme_start_request ( PANID, RF_CHANNEL,
                             BEACON_ORDER, SUPERFRAME_ORDER,
                             true, false, false, false);
    SET_STATE(PEND_START);
    return;
}
```
****

Code Snippet 5.12: An active scan of the 2.4-GHz channel range will be performed, which really gives us nothing to work with as the RF_CHANNEL is set to 11 and the 2.4-GHz frequency band is defined in the Custom Compilation Options window of WinAVR. As you can see, we simply start a PAN following the scan.

Since we have previously defined the operating frequency band (2.4 GHz) and the channel (11) we want to operate on in the WinAVR Custom Compilation Options window, we can throw the rest of our definitions into the

wpan_mlme_start_request function's argument list and start a new PAN. The execution of the wpan_mlme_start_request function generates the callback function usr_mlme_start_conf. If we pretend that MAC_SUCCESS was returned by the wpan_mlme_start_request function in Code Snippet 5.13, the wheels start turning to allow association with the newly crowned PAN Coordinator.

Code Snippet 5.13
****

```
void usr_mlme_start_conf(uint8_t status)
{
    if  (c_status.state == PEND_START)
    {
        if (status == MAC_SUCCESS)
        {
            mac_set_assoc_permit( 1 );
        }
        else
        {
            mac_start_pan();
        }
    }
    return;
}

void mac_set_assoc_permit(uint8_t permit)
{
    wpan_mlme_set_request (macAssociationPermit, &permit, sizeof(permit));
    SET_STATE( PEND_ASSOC_PERMIT );
}

void usr_mlme_set_conf(uint8_t status, uint8_t PIBAttribute )
{
    switch (c_status.state)
    {
        case PEND_SET_SHORT_ADDR:
            if ((status == MAC_SUCCESS) && (PIBAttribute == macShortAddress))
            {
                mac_active_scan();
            }
            break;
        case PEND_ASSOC_PERMIT:
            if ((status == MAC_SUCCESS) && (PIBAttribute == macAssociationPermit))
            {
                SET_STATE( RUN );
                PORTE = 0xff; /* LED's off if we come to here */
            }
```

```
        default:
            break;
    }

    return;
}
```
****

Code Snippet 5.13: Just follow the chain. The wpan_mlme_set_request enables end device association with the PAN Coordinator, which transitions the PAN Coordinator into the RUN state. To give the user a visual that things are good up to this point, the LEDs turned on at the beginning of the PAN creation process are extinguished.

The newly created PAN is ready for work at this point. The PAN Coordinator will constantly be on the lookout for end devices to join its PAN. Nothing useful that the Daintree Networks SNA can pull from the air from our new PAN has been transmitted yet. In fact, nothing will happen that we can capture and interpret with the Daintree Networks SNA application until an end device requests to be associated. And, the only way an end device can get to that point is to run its application. So, let's let the new PAN Coordinator spin and pick up with the end device application code in Code Snippet 5.14.

An AT86RF230 End Device Application

Right now, a new PAN Coordinator is waiting for an end device to request an association and join the PAN. If you're wondering what the end device is up to, you won't be wondering long.

The Includes area of Code Snippet 5.14 is no surprise as the end device will eventually become part of an IEEE 802.15.4 network. The IEEE 802.15.4 network that our end device will attempt to join resides in the 2.4-GHz ISM band. So, there's no reason for the end device to scan the 868-MHz and 915-MHz channels. That's why you see the 11 least-significant bits, which represent 868-MHz and 915-MHz channels, zeroed out in the Macros area ALL_HIGH_BAND_CHANNELS definition.

The Typedefs area needs little explanation as it parallels the PAN Coordinator's Typedef definitions. If you really take your time and go through the device_status_t data structure, you'll see that many of the data structure elements are provided by the PAN Coordinator during association. Don't worry—if that's not clear to you now, the fog will lift when you see the association debug captures.

Code Snippet 5.14
****

```
/*===Includes========================================= */
#include <string.h>
#include <stdint.h>
#include "wpan_defines.h"
#include "ieee_const.h"
#include "ieee_types.h"
#include "wpan.h"
```

```c
/* === Macros ======================================== */
#ifndef RF_CHANNEL
# define RF_CHANNEL (16)
# warning "RF channel undefined, setting to 16"
#endif
#   define _BV(x) (1<<(x))
#define ALL_HIGH_BAND_CHANNELS ((uint32_t)0x07FFF800)
#define CHANNELMASK(a) (1UL<<(a))
#define SCAN_DURATION   (3) /* scan for 3 * symbol period  = 48 µs */

/* macro stores the state value and sets the state led to 0 */
#define SET_STATE(x) do { d_status.state=(x); \
                          PORTE |= (_BV((uint8_t)(x)));} while(0)

/* === Typedefs ======================================== */
typedef enum
{
    INIT_DONE,
    PEND_RESET,
    PEND_SCAN,
    PEND_ASSOCIATE,
    PEND_SET_SHORT_ADDR,
    PEND_START,
    RUN,
} device_state_t;

typedef struct
{
    bool led;
    bool switch_pressed;
    uint16_t device_short_address;
    uint8_t  coord_address_mode;
    uint64_t coord_address;
    uint16_t pan_id;
    uint8_t logical_channel;
    uint8_t msdu_handle;
    device_state_t state;
} device_status_t;


/* === Static Variables ======================================== */
static device_status_t d_status;

/* === Prototypes ======================================== */
static void application_init(void);
static void switch_task(void);
void mac_do_reset(void);
static void mac_scan(void);
static void mac_associate(void);

/* === Implementation ======================================== */
int main(void)
```

```
{
    application_init();

    mac_do_reset();

    while(1)
    {
        while(wpan_task())
        {
            /* only short running tasks are called here */
        }
        /* main user task */
        switch_task();
    }
}


static void application_init(void)
{
    /* reset global application status variable */
    memset(&d_status, 0, sizeof(d_status));

    /* init IO ports */
    DDRE = 0xDF;  /* PE5 is input */
    PORTE = 0x00;  /* switch all leds ON (inverse logic) */

    /* init mac layer */
    wpan_init();
    SET_STATE(INIT_DONE);

    /* enable interrupts */
    sei();
    return;
}


static void switch_task(void)
{
    if (d_status.state == RUN)
    {
        bool send_data = false;

        if (!d_status.switch_pressed)
        {
            /* check if button is pressed. */
            if ((PINE & 0x20) == 0x00)
            {
                d_status.switch_pressed = true;
                d_status.led = !d_status.led;
                send_data = true;
            }
```

```
            }
            else
            {
                /* check if button is released. */
                if ((PINE & 0x20) == 0x20)
                {
                    d_status.switch_pressed = false;
                }
            }

            if (send_data)
            {
                /* send data */
                wpan_mcpsdata_addr_t addr_info;
                addr_info.SrcAddrMode = WPAN_ADDRMODE_SHORT;
                addr_info.SrcPANId = d_status.pan_id;
                addr_info.SrcAddr = d_status.device_short_address;
                addr_info.DstAddrMode = d_status.coord_address_mode;
                addr_info.DstPANId = d_status.pan_id;
                addr_info.DstAddr = d_status.coord_address;

                wpan_mcps_data_request(&addr_info,
                                d_status.msdu_handle++, WPAN_TXOPT_ACK,
                                (void *)&d_status.led, sizeof(uint8_t));
            }
        }
    return;
}

void mac_do_reset()
{
    wpan_mlme_reset_request( true );
    SET_STATE( PEND_RESET );
}

void usr_mlme_reset_conf (uint8_t status)
{
    if ((status == MAC_SUCCESS) && (d_status.state == PEND_RESET))
    {
        mac_scan();
    }

    return;
}


static void mac_scan(void)
{
    uint32_t chanmsk;
    chanmsk = CHANNELMASK(RF_CHANNEL);
    wpan_mlme_scan_request(MLME_SCAN_TYPE_ACTIVE, chanmsk, SCAN_DURATION);
```

```
    SET_STATE( PEND_SCAN );
    return;
}


void usr_mlme_scan_conf(uint8_t status, uint8_t ScanType,
                        uint32_t UnscannedChannels, uint8_t ResultListSize,
                        uint8_t *data, uint8_t data_length)
{
    bool scan_success = false;

    if ((status == MAC_SUCCESS) && (d_status.state == PEND_SCAN))
    {
        /* there should only be one PAN descriptor */
        if (ResultListSize == 1)
        {
            scan_success = true;
            pandescriptor_long_t *pandesc = (pandescriptor_long_t *)data;

            /* save information from the PAN Descriptor */
            d_status.coord_address_mode = pandesc->CoordAddrMode;
            d_status.coord_address = pandesc->CoordAddress;
            d_status.pan_id = pandesc->CoordPANId;
            d_status.logical_channel = pandesc->LogicalChannel;

            /* associate to the PAN Coordinator */
            mac_associate();
        }
    }

    if (!scan_success)
    {
        /* no success, scan again */
        mac_scan();
    }

    return;
}


static void mac_associate(void)
{
    uint8_t capability_info;
    capability_info = WPAN_CAP_FFD | WPAN_CAP_PWRSOURCE |\
                      WPAN_CAP_RXONWHENIDLE | WPAN_CAP_ALLOCADDRESS;

    wpan_mlme_associate_request(d_status.logical_channel,
                                d_status.coord_address_mode,
                                d_status.pan_id, d_status.coord_address,
                                capability_info, false);
```

```
        SET_STATE( PEND_ASSOCIATE );
        return;
}


void usr_mlme_associate_conf ( uint16_t AssocShortAddress, uint8_t status)
{
    if ((status == MAC_SUCCESS) && (d_status.state == PEND_ASSOCIATE))
    {
        /* save the device short address */
        d_status.device_short_address = AssocShortAddress;

        /* mark that association is complete */
        SET_STATE( RUN );

        /* turn all LEDs off after successfull association */
        PORTE = 0xFF;
    }
    else
    {
        /* somethig went wrong, try  association again */
        mac_associate();
    }

    return;
}


void usr_mcps_data_ind(wpan_mcpsdata_addr_t *pAddrInfo,
                       uint8_t mpduLinkQuality, uint8_t SecurityUse,
                       uint8_t ACLEntry, uint8_t msduLength, uint8_t *msdu)
{
    if ((d_status.state == RUN) && (pAddrInfo->DstPANId == d_status.pan_id))
    {
        /* Data packet contains LED information. */
        PORTE = ~((uint8_t)*msdu);
    }
}
```
****
Code Snippet 5.14: Here's your proof that the Atmel IEEE 802.15.4 MAC can be applied to an FFD or an RFD as the same Atmel IEEE 802.15.4 MAC API calls used in the PAN Coordinator application are used here in the end device application.

The end device application looks a lot like the PAN Coordinator application until we get to the wpan_mlme_reset_request's callback function. The code within the usr_mlme_reset_conf callback function in Code Snippet 5.15 lets the end-device dog go huntin'.

Code Snippet 5.15

```
**************************************************************************
void mac_do_reset()
{
    wpan_mlme_reset_request( true );
    SET_STATE( PEND_RESET );
}

void usr_mlme_reset_conf (uint8_t status)
{
    if ((status == MAC_SUCCESS) && (d_status.state == PEND_RESET))
    {
        mac_scan();
    }

    return;
}

static void mac_scan(void)
{
    uint32_t chanmsk;
    chanmsk = CHANNELMASK(RF_CHANNEL);
    wpan_mlme_scan_request(MLME_SCAN_TYPE_ACTIVE, chanmsk, SCAN_DURATION);
    SET_STATE( PEND_SCAN );
    return;
}
**************************************************************************
```

Code Snippet 5.15: The end-device application is hardcoded to only scan channel 11 as RF_CHANNEL has been predefined in WinAVR as 11.

I'm sure you can follow the logical flow in Code Snippet 5.15. The resulting action carried out in Code Snippet 5.15 is the transmission of a Beacon Request message on channel 11. The Beacon Request message is transmitted as a broadcast and any PAN Coordinator in earshot that is permitting association will answer with a Beacon frame describing the PAN it is in charge of. The Beacon Request frame that the end device generated is shown in Sniffer Capture 5.1.

Sniffer Capture 5.1

```
**************************************************************************
Frame 1 (Length = 10 bytes)
        Time Stamp: 11:01:25.000
        Frame Length: 10 bytes
        Capture Length: 10 bytes
        Link Quality Indication: 116
IEEE 802.15.4
        Frame Control: 0x0803
                .... .... .... .011 = Frame Type: Command (0x0003)
                .... .... .... 0... = Security Enabled: Disabled
                .... .... ...0 .... = Frame Pending: No more data
```

```
                   .... .... ..0. ....  = Acknowledgment Request: Acknowledgment
not required
                   .... .... .0.. ....  = Intra PAN: Not within the PAN
                   .... ..00 0... ....  = Reserved
                   .... 10.. .... ....  = Destination Addressing Mode: Address
field contains a 16-bit short address (0x0002)
                   ..00 .... .... ....  = Reserved
                   00.. .... .... ....  = Source Addressing Mode: PAN identifier
and address field are not present (0x0000)
        Sequence Number: 144
        Destination PAN Identifier: 0xffff
        Destination Address: 0xffff
        MAC Payload
             Command Frame Identifier = Beacon Request: (0x07)
        Frame Check Sequence: Correct

0000:   03 08 90 ff ff ff ff 07 .. ..                    .........
*************************************************************************
```

Sniffer Capture 5.1: This is akin to putting your message in a bottle and casting it out into the ocean. However, the odds of getting a positive response from the Beacon Request are much greater than getting a return note in a bottle on the beach.

The PAN Coordinator fired off a Beacon in response to the Beacon Request from the end device. Sniffer Capture 5.2 holds the details contained within the Beacon that was transmitted by the PAN Coordinator.

Sniffer Capture 5.2
```
*************************************************************************
Frame 2 (Length = 13 bytes)
        Time Stamp: 11:01:25.007
        Frame Length: 13 bytes
        Capture Length: 13 bytes
        Link Quality Indication: 208
IEEE 802.15.4
        Frame Control: 0x8000
                   .... .... .... .000  = Frame Type: Beacon (0x0000)
                   .... .... .... 0...  = Security Enabled: Disabled
                   .... .... ...0 ....  = Frame Pending: No more data
                   .... .... ..0. ....  = Acknowledgment Request: Acknowledgment
not required
                   .... .... .0.. ....  = Intra PAN: Not within the PAN
                   .... ..00 0... ....  = Reserved
                   .... 00.. .... ....  = Destination Addressing Mode: PAN
identifier and address field are not present (0x0000)
                   ..00 .... .... ....  = Reserved
                   10.. .... .... ....  = Source Addressing Mode: Address field
contains a 16-bit short address (0x0002)
        Sequence Number: 79
        Source PAN Identifier: 0xcafe
        Source Address: 0xbabe
        MAC Payload
```

```
          Superframe Specification: 0xcfff
                 .... .... .... 1111  = Beacon Order (0x000f)
                 .... .... 1111 ....  = Superframe Order (0x000f)
                 .... 1111 .... ....  = Final CAP Slot (0x000f)
                 ...0 .... .... ....  = Battery Life Extension: Disabled
                 ..0. .... .... ....  = Reserved
                 .1.. .... .... ....  = PAN Coordinator: Transmitter is
a PAN Coordinator
                 1... .... .... ....  = Association Permit: Coordinator
accepting Association Requests
          GTS Specification: 0x00
                 .... .000  = GTS Descriptor Count (0x00)
                 .000 0...  = Reserved
                 0... ....  = GTS Permit: Coordinator not accepting GTS
Requests
          Pending Address Specification: 0x00
                 .... .000  = Number of short Addresses pending: 0
                 .... 0...  = Reserved
                 .000 ....  = Number of extended Addresses pending: 0
                 0... ....  = Reserved
     Frame Check Sequence: Correct

0000:   00 80 4f fe ca be ba ff cf 00 00 .. ..        ..O~J>:.O....
```

Sniffer Capture 5.2: This collection of data tells the requesting device how to contact the PAN Coordinator, what superframe logic to use, what addressing mode to use and if the PAN Coordinator is accepting new applicants to its PAN via association.

As we have come to expect, the result of the channel 11 scan request is returned in a confirmation message, whose code is shown in Code Snippet 5.16. The data returned to the end device in the identifying Beacon is used to populate the confirmation fields.

The end device in this situation has been limited to the single PAN we just spawned. So, the only Beacon response was generated by the newly created PAN Coordinator. Since the end device is destined to join our new PAN, it had better collect enough information to be able to get an association ticket on the Beaconing PAN. So, the code in Code Snippet 5.16 collects important information from the PAN Descriptor, which was transmitted inside the Beacon.

Code Snippet 5.16

```
void usr_mlme_scan_conf(uint8_t status, uint8_t ScanType,
                    uint32_t UnscannedChannels, uint8_t ResultListSize,
                    uint8_t *data, uint8_t data_length)
{
    bool scan_success = false;

    if ((status == MAC_SUCCESS) && (d_status.state == PEND_SCAN))
    {
        /* there should only be one PAN descriptor */
        if (ResultListSize == 1)
```

```
            {
                scan_success = true;
                pandescriptor_long_t *pandesc = (pandescriptor_long_t *)data;

                /* save information from the PAN Descriptor */
                d_status.coord_address_mode = pandesc->CoordAddrMode;
                d_status.coord_address = pandesc->CoordAddress;
                d_status.pan_id = pandesc->CoordPANId;
                d_status.logical_channel = pandesc->LogicalChannel;

                /* associate to the PAN Coordinator */
                mac_associate();
            }
        }

        if (!scan_success)
        {
            /* no success, scan again */
            mac_scan();
        }

        return;
}
```
****

Code Snippet 5.16: The ResultListSize value reflects the number of PAN Coordinators that responded to the end device's Beacon Request. Note that after the end device gleans the information it needs for association, it attempts to associate with the PAN Coordinator that transmitted the answering Beacon.

I stopped the execution of the end-device code just before the end device attempted to associate with the PAN. I figured some of you would like to see how the data looks from the ATmega1281's point of view. The contents of the d_status data structure lying within the ATmega1281's SRAM are depicted in Screen Capture 5.1.

The end device now has enough real information to request association with the PAN that it accepted the Beacon from. The necessary MLME primitive is passed in Code Snippet 5.17.

Code Snippet 5.17
****

```
static void mac_associate(void)
{
    uint8_t capability_info;
    capability_info = WPAN_CAP_FFD | WPAN_CAP_PWRSOURCE |\
                      WPAN_CAP_RXONWHENIDLE | WPAN_CAP_ALLOCADDRESS;

    wpan_mlme_associate_request(d_status.logical_channel,
                                d_status.coord_address_mode,
                                d_status.pan_id, d_status.coord_address,
                                capability_info, false);
```

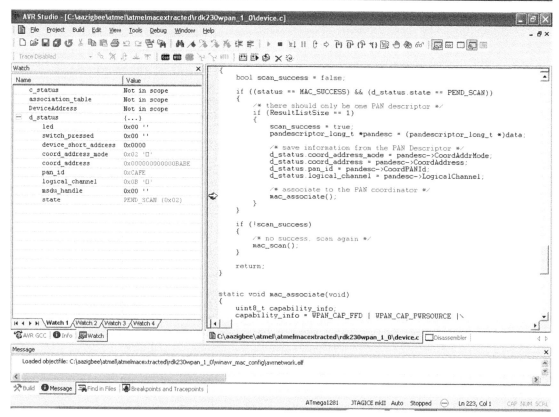

Screen Capture 5.1: The idea behind showing you this shot is to provide a reference as to what is going on inside the ATmega1281 as it relates to what the end device code is doing.

```
    SET_STATE( PEND_ASSOCIATE );
    return;
}
```
****

Code Snippet 5.17: Note that, in addition to the standard association request mumbo jumbo, the end device is also passing along its capabilites to the PAN Coordinator. The ATmega1281 supporting the AT86RF230 gives the end device enough compute power to be an FFD if the application calls for it.

Sniffer Capture 5.3
****
```
Frame 3 (Length = 21 bytes)
        Time Stamp: 11:01:25.146
        Frame Length: 21 bytes
```

```
        Capture Length: 21 bytes
        Link Quality Indication: 128
IEEE 802.15.4
        Frame Control: 0xc823
                 .... .... .... .011  = Frame Type: Command (0x0003)
                 .... .... .... 0...  = Security Enabled: Disabled
                 .... .... ...0 ....  = Frame Pending: No more data
                 .... .... ..1. ....  = Acknowledgment Request: Acknowledgment
required
                 .... .... .0.. ....  = Intra PAN: Not within the PAN
                 .... ..00 0... ....  = Reserved
                 .... 10.. .... ....  = Destination Addressing Mode: Address
field contains a 16-bit short address (0x0002)
                 ..00 .... .... ....  = Reserved
                 11.. .... .... ....  = Source Addressing Mode: Address field
contains a 64-bit extended address (0x0003)
        Sequence Number: 145
        Destination PAN Identifier: 0xcafe
        Destination Address: 0xbabe
        Source PAN Identifier: 0xffff
        Source Address: 0x000425ffff170537
        MAC Payload
                Command Frame Identifier = Association Request: (0x01)
                Capability Information: 0x8e
                     .... ...0  = Alternate PAN Coordinator: Not capable of
becoming PAN Coordinator
                     .... ..1.  = Device Type: FFD
                     .... .1..  = Power Source: Receiving power from
alternating current mains
                     .... 1...  = Receiver on when idle: Enables receiver
when idle
                     ..00 ....  = Reserved
                     .0.. ....  = Security Capability: Not capable of using
security suite
                     1... ....  = Allocate Address: Coordinator should
allocate short address
        Frame Check Sequence: Correct


0000:    23 c8 91 fe ca be ba ff ff 37 05 17 ff ff 25 04    #H.~J>:..7....%.
0010:    00 01 8e .. ..                                     .....
************************************************************************
```

Sniffer Capture 5.3: This marks the first time we've seen the 64-bit IEEE address actually utilized in an IEEE 802.15.4 transaction. The Daintree Networks SNA application makes it all look so easy, as you can gain an understanding of what is happening by simply reading through the Daintree Networks SNA trace text.

An acknowledgment has been requested (see Screen Capture 5.3 Frame Control) and the PAN Coordinator replies in kind in Sniffer Capture 5.4.

Sniffer Capture 5.4

```
*******************************************************************************
Frame 4 (Length = 5 bytes)
       Time Stamp: 11:01:25.147
       Frame Length: 5 bytes
       Capture Length: 5 bytes
       Link Quality Indication: 208
IEEE 802.15.4
       Frame Control: 0x0002
                  .... .... .... .010 = Frame Type: Acknowledgment (0x0002)
                  .... .... .... 0... = Security Enabled: Disabled
                  .... .... ...0 .... = Frame Pending: No more data
                  .... .... ..0. .... = Acknowledgment Request: Acknowledgment
not required
                  .... .... .0.. .... = Intra PAN: Not within the PAN
                  .... ..00 0... .... = Reserved
                  .... 00.. .... .... = Destination Addressing Mode: PAN
identifier and address field are not present (0x0000)
                  ..00 .... .... .... = Reserved
                  00.. .... .... .... = Source Addressing Mode: PAN identifier
and address field are not present (0x0000)
       Sequence Number: 145
       Frame Check Sequence: Correct

0000:    02 00 91 .. ..                                    .....
*******************************************************************************
```

Sniffer Capture 5.4: Recall that the Sequence Number is the key here. This acknowledgment pairs with the association request message in Sniffer Capture 5.3.

According to the IEEE 802.15.4 specification, if the end device doesn't get a warm fuzzy about its association request within a specified response time, the end device will issue a MLME-ASSOCIATE.confirm primitive with a status of NO_DATA. I'd say the end device did just that in Sniffer Capture 5.5.

Sniffer Capture 5.5

```
*******************************************************************************
Frame 5 (Length = 16 bytes)
       Time Stamp: 11:01:25.644
       Frame Length: 16 bytes
       Capture Length: 16 bytes
       Link Quality Indication: 112
IEEE 802.15.4
       Frame Control: 0xc023
                  .... .... .... .011 = Frame Type: Command (0x0003)
                  .... .... .... 0... = Security Enabled: Disabled
                  .... .... ...0 .... = Frame Pending: No more data
                  .... .... ..1. .... = Acknowledgment Request: Acknowledgment
required
                  .... .... .0.. .... = Intra PAN: Not within the PAN
                  .... ..00 0... .... = Reserved
```

```
                    .... 00.. .... ....    = Destination Addressing Mode: PAN
identifier and address field are not present (0x0000)
                    ..00 .... .... ....    = Reserved
                    11.. .... .... ....    = Source Addressing Mode: Address field
contains a 64-bit extended address (0x0003)
        Sequence Number: 146
        Source PAN Identifier: 0xcafe
        Source Address: 0x000425ffff170537
        MAC Payload
                Command Frame Identifier = Data Request: (0x04)
        Frame Check Sequence: Correct


0000:   23 c0 92 fe ca 37 05 17 ff ff 25 04 00 04 .. ..    #@.~J7....%.....
*****************************************************************************
```

Sniffer Capture 5.5: Impatient little bugger, isn't it? Note the use of the end device's 64-bit IEEE address here and the assumption of the PAN Identifier which the end device has not yet associated with. The PAN Identifier is a must here as that is the only way for the requesting end device to address the PAN at this point in time.

An acknowledgment with a Sequence Number of 146 is offered up by the PAN Coordinator in Sniffer Capture 5.6 as requested in Sniffer Capture 5.5.

Sniffer Capture 5.6

```
*****************************************************************************
Frame 6 (Length = 5 bytes)
        Time Stamp: 11:01:25.645
        Frame Length: 5 bytes
        Capture Length: 5 bytes
        Link Quality Indication: 208
IEEE 802.15.4
        Frame Control: 0x0012
                    .... .... .... .010    = Frame Type: Acknowledgment (0x0002)
                    .... .... .... 0...    = Security Enabled: Disabled
                    .... .... ...1 ....    = Frame Pending: More data
                    .... .... ..0. ....    = Acknowledgment Request: Acknowledgment
not required
                    .... .... .0.. ....    = Intra PAN: Not within the PAN
                    .... ..00 0... ....    = Reserved
                    .... 00.. .... ....    = Destination Addressing Mode: PAN
identifier and address field are not present (0x0000)
                    ..00 .... .... ....    = Reserved
                    00.. .... .... ....    = Source Addressing Mode: PAN identifier
and address field are not present (0x0000)
        Sequence Number: 146
        Frame Check Sequence: Correct


0000:   12 00 92 .. ..                                      .....
*****************************************************************************
```

Sniffer Capture 5.6: The song remains the same…sorta. The only difference in this acknowledgment and the acknowledgment in Sniffer Capture 5.4 is the Sequence Number.

The PAN Coordinator determines that it does indeed have an available slot for the requesting end device and fires off the association response message in Sniffer Capture 5.7. This is the end device's lucky day. The PAN Coordinator reveals its IEEE 64-bit address (000425ffff170436) and confirms the association of the end device (000425ffff170537) with the PAN.

Sniffer Capture 5.7
****
```
Frame 7 (Length = 29 bytes)
        Time Stamp: 11:01:25.651
        Frame Length: 29 bytes
        Capture Length: 29 bytes
        Link Quality Indication: 208
IEEE 802.15.4
        Frame Control: 0xcc23
                .... .... .... .011  = Frame Type: Command (0x0003)
                .... .... .... 0...  = Security Enabled: Disabled
                .... .... ...0 ....  = Frame Pending: No more data
                .... .... ..1. ....  = Acknowledgment Request: Acknowledgment
required
                .... .... .0.. ....  = Intra PAN: Not within the PAN
                .... ..00 0... ....  = Reserved
                .... 11.. .... ....  = Destination Addressing Mode: Address
field contains a 64-bit extended address (0x0003)
                ..00 .... .... ....  = Reserved
                11.. .... .... ....  = Source Addressing Mode: Address field
contains a 64-bit extended address (0x0003)
        Sequence Number: 173
        Destination PAN Identifier: 0xcafe
        Destination Address: 0x000425ffff170537
        Source PAN Identifier: 0xcafe
        Source Address: 0x000425ffff170436
        MAC Payload
                Command Frame Identifier = Association Response: (0x02)
                Short Address: 0x0000
                Association Status: Association Successful (0x00)
        Frame Check Sequence: Correct

0000:   23 cc ad fe ca 37 05 17 ff ff 25 04 00 fe ca 36    #L-~J7....%..~J6
0010:   04 17 ff ff 25 04 00 02 00 00 00 .. ..              ....%........
```
****
Sniffer Capture 5.7: Note the choice to use the official IEEE 64-bit addressing scheme. Each of the Z-Link modules is tagged with the 64-bit address and using the 64-bit addressing scheme makes it a bit easier to keep up with which Z-Link module is doing what.

In the meantime, the PAN Coordinator is logging the association of the new end device as shown in Screen Capture 5.2. The PAN Coordinator also issues an acknowledgment with a Sequence Number of 173.

Screen Capture 5.2: The end device that just associated is logged into slot 0 (zero) of the association_ table using the end device's IEEE 64-bit address.

Note that in Code Snippet 5.18 the short address is retained. That's because the short addressing mode will be used in the actual application, which is posted in Code Snippet 5.19.

Code Snippet 5.18
****
```
void usr_mlme_associate_conf ( uint16_t AssocShortAddress, uint8_t status)
{
    if ((status == MAC_SUCCESS) && (d_status.state == PEND_ASSOCIATE))
    {
        /* save the device short address */
        d_status.device_short_address = AssocShortAddress;

        /* mark that association is complete */
        SET_STATE( RUN );

        /* turn all LEDs off after successful association */
        PORTE = 0xFF;
    }
    else
    {
        /* somethig went wrong, try  association again */
        mac_associate();
    }

    return;
}
```
****
Code Snippet 5.18: When the LEDs turn off on the end device, things are good. Note that a short address is also assigned to the end device at association time. We've already discussed the inefficiencies of using the IEEE 64-bit address in our limited-space data transactions.

Unfortunately, I can't show you the LEDs illuminating and extinguishing. However, the application code in Code Snippet 5.19, which is running on the end device, is very easy to follow.

Code Snippet 5.19
****
```
static void switch_task(void)
{
    if (d_status.state == RUN)
    {
        bool send_data = false;

        if (!d_status.switch_pressed)
        {
            /* check if button is pressed. */
            if ((PINE & 0x20) == 0x00)
```

```
    {
        d_status.switch_pressed = true;
        d_status.led = !d_status.led;
        send_data = true;
    }
}
else
{
    /* check if button is released. */
    if ((PINE & 0x20) == 0x20)
    {
        d_status.switch_pressed = false;
    }
}

if (send_data)
{
    /* send data */
    wpan_mcpsdata_addr_t addr_info;
    addr_info.SrcAddrMode = WPAN_ADDRMODE_SHORT;
    addr_info.SrcPANId = d_status.pan_id;
    addr_info.SrcAddr = d_status.device_short_address;
    addr_info.DstAddrMode = d_status.coord_address_mode;
    addr_info.DstPANId = d_status.pan_id;
    addr_info.DstAddr = d_status.coord_address;

    wpan_mcps_data_request(&addr_info,
                    d_status.msdu_handle++, WPAN_TXOPT_ACK,
                    (void *)&d_status.led, sizeof(uint8_t));
    }
}
return;
}
```

****

Code Snippet 5.19: The passing of the MCPS Data Request primitive kicks off the transmission of the switch status.

Thanks to the folks at Atmel, you should have a pretty good idea about how IEEE 802.15.4 can be put to work for you using an Atmel AVR microcontroller. We've covered a lot of ground in this chapter. So, let's recap what happened from the PAN Coordinator's point of view:

- SET MAC SHORT ADDRESS
- wpan_mlme_set_request
- usr_mlme_set_conf
- SCAN FOR LOWEST INACTIVE CHANNEL
- mac_active_scan
- wpan_mlme_scan_request

- usr_mlme_scan_conf
- START NEW PAN ON CHOSEN CHANNEL
- mac_start_pan()
- wpan_mlme_start_request
- usr_mlme_start_conf
- PERMIT END DEVICES TO ASSOCIATE WITH NEW PAN
- mac_set_assoc_permit(1)
- wpan_mlme_set_request
- usr_mlme_set_conf
- NEW PAN ESTABLISHED WAITING FOR ASSOCIATION REQUESTS
- State machine state is set to RUN using SET_STATE(RUN)
- AN END DEVICE REQUESTS ASSOCIATION
- usr_mlme_associate_ind
- REGISTER END DEVICE BY IEEE LONG ADDRESS
- mac_register_device
- POSITIVE Acknowledgment TO ASSOCIATED END DEVICE
- wpan_mlme_associate_response
- WAIT FOR END DEVICE TO SEND DATA (BUTTON PUSH)
- usr_mcps_data_ind
- UPDATE LED PORT AND SEND BUTTON DATA TO ALL ASSOCIATED END DEVICES
- mac_send_data
- REPEAT PROCESS FROM usr_mcps_data_ind

Now, let's compress the events as seen from the end device's point of view:

- SCAN FOR A PAN COORDINATOR
- wpan_mlme_scan_request
- GATHER SCAN INFORMATION
- usr_mlme_scan_conf
- REQUEST ASSOCIATION TO PREFERRED PAN
- wpan_mlme_associate_request

- ASSOCIATE AND RECEIVE SHORT ADDRESS

- usr_mlme_associate_conf

- EXECUTE MAIN TASK

- wpan_mcps_data_request

- REFLECT STATUS OF LEDs FROM RECEIVED DATA

- usr_mcps_data_ind

- REPEAT PROCESS FROM EXECUTE MAIN TASK

Yet One More Way

In this chapter I've presented to you the Atmel flavor of IEEE 802.15.4. You can download the Atmel IEEE 802.15.4 MAC we've studied in this chapter from the Atmel website at *www. atmel.com.*

About Atmel

Founded in 1984, Atmel is a worldwide leader in the design and manufacture of micro-controllers, advanced logic, mixed-signal, nonvolatile memory and radio frequency (RF) components. Atmel microcontrollers have also been the targets of some of my magazine columns.

The Guy Lombardo of Halloween? Here's a hint:

They did the Mash.

Bobby "Boris" Pickett, other than being known as the Guy Lombardo of Halloween, is the only active performer whose original recording charted in the top 100 three times. You know the song—it's called "Monster Mash."

Let's move on to the next brick in the ZigBee/IEEE 802.15.4 wall and take a look at the Texas Instruments/Chipcon art of IEEE 802.15.4. However, before you turn the page, a beautiful teenager named Arlene should have kept her man when she had the chance instead of laughing at him while swooning over an Elvis song. I literally grew up with this guy and so did most of America. Who is he?

They Do Everything BIG in Texas

Every ZigBee network has to ride on an IEEE 802.15.4 radio. Thus far, you've seen how Atmel does IEEE 802.15.4 transceivers with their AT86RF230. You've also ridden with me in the ZMD Cadillac, the ZMD44102, and while your hair was blowing in the wind (I have no hair on my head, thank you) we saw how the ZMD folks do IEEE 802.15.4 radios. Here we are and the clock is tickin'. So, let's take a look at how the folks at Texas Instruments/Chipcon transmit those IEEE 802.15.4 chips.

One of Two

Well, in our little world this is one of two. The first Texas Instruments/Chipcon transceiver IC we will examine is the CC2420. Thank goodness for specifications. At least, thank goodness for specifications that are adhered to by those that really must follow their threads for reasons of compatibility. The Texas Instruments/Chipcon CC2420 follows in the path of all of the other IEEE 802.15.4/ZigBee-ready transceivers we've experienced up to now. It will be of no surprise to you that the CC2420 is a single-chip 2.4-GHz IEEE 802.15.4-compliant RF transceiver. I could go down the list calling out and explaining I/Q-based receivers and transmitters, 128-byte receive and transmit data buffering, etc., and it would all sound the same. And, in a sense, it would be all the same as the end result for all IEEE 802.15.4 radios is the same. The good news is that no two IEEE 802.15.4 radios from differing manufacturers will logically look exactly alike nor will they contain identical register sets and special features. Another differentiation factor lies in the support packages that stand behind a transceiver IC. In the case of Texas Instruments/Chipcon, the support packages are really top notch. The test and prototyping/debugging hardware platforms from Texas Instruments/Chipcon aren't too shabby either.

Let's begin by looking closely at the actual CC2420 hardware. So, please turn your attention to Figure 6.1, which is an abbreviated block diagram representing the CC2420.

Let's take a walk down the CC2420's receive path first. The incoming RF signal is amplified by the LNA (low noise amplifier) and down-converted in the quadrature (remember I and Q?) stage. The down-converted 2-MHz IF I/Q signal is passed through a bandpass filter and amplified before being digitized by an analog-to-digital converter. When the digitized signal contents leave the analog-to-digital converter, all of the necessary magic is performed that turns those chips into bytes that can be recognized as data by the host microcontroller.

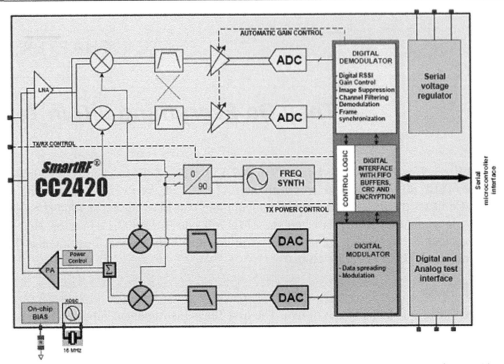

Figure 6.1: Abbreviated block diagram of CC2420. Trust me. The more you see figures like this, the more you will appreciate them, as because of them, IEEE 802.15.4 radio concepts will become second nature to you.

The CC2420 has an SFD (Start of Frame Delimiter) pin that goes high when an SFD has been detected. Incoming data is then piped into the CC2420's 128-byte receive FIFO. Just like the previous IEEE 802.15.4 radios we've peeked at, the CC2420 silicon performs the incoming CRC check. The CC2420 also sports a CCA (Clear Channel Assessment) pin that is available to the host microcontroller during receive mode.

Taking the CC2420 internal components within Figure 6.1 left to right, we begin with hex data contained within the CC2420's 128-byte transmit FIFO being dusted with goo that turns the hex bytes into chips. The CC2420 transmit circuitry generates the preamble and SFD, which, as you already know, must go out before the data. Meanwhile, the chips are converted to analog signals and fed into an analog lowpass filter. The output of the analog lowpass filter is fed into the all-familiar I/Q quadrature mixer stage and is up-converted. The resultant up-converted analog signal is then fed to the CC2420's PA (power amplifier), which pitches the whole mess out of its differential RF interface and into the antenna. The CC2420's internal TX/RX control circuitry uses the TXRX_SWITCH, RF_P and RF_N pins to provide bias to the PA and LNA for automatic switching of the antenna between the PA and LNA inputs.

A crystal is required to provide the reference frequency for the CC2420's frequency synthesizer. Power for the CC2420's internals is provided by an on-chip 1.8V voltage regulator, which can be enabled or disabled via the CC2420's VREG_EN digital interface pin. Keeping with IEEE 802.15.4 and ZigBee's low-power motto, the CC2420 also incorporates an on-chip battery monitor if you choose to employ it.

The CC2420's differential RF I/O interface is high impedance. A common way of terminating the CC2420's 50Ω RF interface is shown schematically in Figure 6.2. All of the components from the antenna back to the RF_P, RF_N and TXRX_SWITCH pins make up what is called a discrete balun.

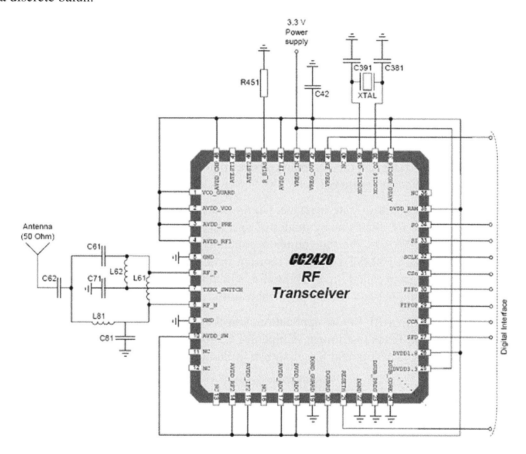

Figure 6.2: I don't wear a pointy hat with stars and moons on it. So, don't expect a big dissertation on the discrete balun. If you're considering building a CC2420 radio from scratch, save yourself the trouble as Texas Instruments/Chipcon recommends you use their layout, which is provided free for all.

Everything you and I will do that will directly affect the CC2420 will be done via the CC2420's SPI portal. The CC2420 is configured as an SPI slave. CC2420 configuration information and buffered data flow make up the traffic flowing across the CC2420's SPI portal. The CC2420 contains 33 16-bit configuration and status registers, 15 command strobe registers, and a pair of 8-bit FIFO access registers.

A register read or write is a 24-bit transaction. Standard SPI protocol is employed with the CC2420's CSn (Chip Select) pin kept logically low during the SPI transfer. The first eight bits of a CC2420 SPI transaction consist of a RAM/Register select bit followed by a R/W (read/write) bit and six bits of address information. Sixteen bits of data follows.

In addition to stuff we load into the CC2420's configuration registers and RAM, the CC2420 also responds to command strobes. A command strobe is a single-byte instruction that is executed by the CC2420. Command strobes do things like enable the crystal oscillator or enable the receive mode. There are 15 command strobes.

The function of a command strobe is activated when the desired command strobe register is accessed with a register-write operation. The exception is that no data is transferred to the command-strobe register. The RAM/Register and R/W bits set to 0 (zero) and the six command strobe address bits (ranging from 0x00 to 0x0E) are all that flow during a command-strobe register-write operation.

The CC2420 works around radio states in a similar manner to the Atmel AT86RF230. The flow of CC2420 radio states and the command strobes that interact with the CC2420 radio states are laid out for you in Figure 6.3.

The CC2420 contains a built-in state machine. Command strobes and internal events create movement between the CC2420's many modes of operation. To assist in your understanding of just how the CC2420 works, the state numbers contained within brackets in Figure 6.3 are reflected in the CC2420's FSMSTATE status register. In addition, Texas Instruments/Chipcon supplies a support program called SmartRF Studio, which will allow us to traverse the state diagram shown in Figure 6.3 and see the physical results.

Before we fire up the SmartRF Studio application, you'll need to see what we're using for CC2420 hardware. The Texas Instruments/Chipcon CC2400EB Evaluation Board, which is part of the CC2420DK Development Kit, is shown in Photo 6.1.

Photo 6.1: There's just enough hardware here to allow us to manipulate the CC2420 innards. Note the absence of a recognizable microcontroller. The interface smarts are all contained within a Xilinx SPARTAN FGPA.

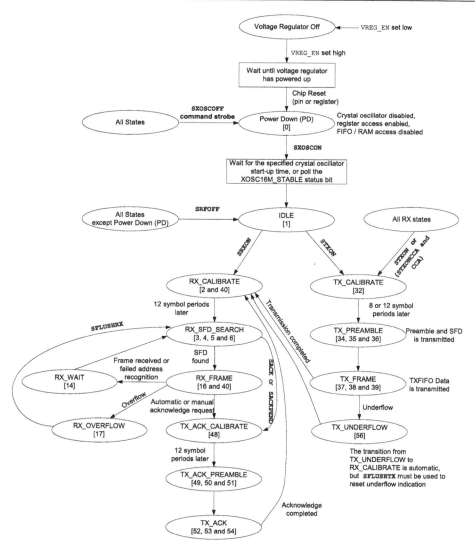

Figure 6.3: Nothing happens here without having enabled the CC2420's voltage regulator and crystal oscillator. Command strobes such as SRXON and STXON move the CC2420 from an idle state to receive and transmit modes respectively.

The CC2400EB looks pretty busy but we can clear all of that up rather quickly. Starting from the far left, we see RS-232 and USB interface connectors, which are supported by a MAXIM RS-232 converter IC and a Cypress EZ-USB IC. The Xilinx FPGA is supported by an IDT memory module that is stationed directly to the right of the FPGA, which is located at the center of the CC2400EB. Believe it or not, the Cypress EZ-USB microcontroller and the Xilinx FPGA are dumb until the load arrives via the USB connection. Thus, the CC2400EB can't function without being attached to a personal computer. We can release the CC2420EM radio module from the clutches of the FPGA by downloading an included FPGA codeset. Relieving the FPGA of command puts the CC2420EM radio module's digital signals on the Test Port 1 pins. Test Port 2 is in parallel with Test Port 1 to allow the easy connection of logic monitors to the Test Port 2 pin set. Oscilloscopes and signal generators can be attached to the quad of SMA receptacles you see in the bottom right of Photo 6.1.

Remember that discrete balun I refused to expound upon earlier? Well, you can see it close up between the CC2420 and the antenna connector in Photo 6.2.

Photo 6.2: Here's a bird's-eye view of the CC2420EM radio module that you see mounted in its socket on the CC2420EB in Photo 6.1. I've removed the antenna for a clearer shot at the component layout.

While the vision in Photo 6.2 is fresh in your mind (and close by), check out another manufacturer's CC2420 radio in Photo 6.2. Can you see the discrete balun in Photo 6.2? Does anything else look "familiar" between Photos 6.2 and 6.3?

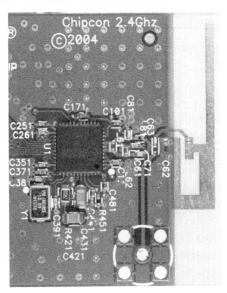

Photo 6.3: This circuit's components and layout are identical to the CC2420 circuit you see in Photo 6.2. Note that you can simply move C63 (to become C62) to use the SMA antenna connector.

There's really very little to talk about when it comes to the CC2420's external circuitry. With the exception of the discrete balun, most of the CC2420's external circuitry consists of bypass capacitors. Admiration time for the CC2420 hardware is over. Let's put all of this pretty CC2420 hardware to work.

As you can see at the bottom of Screen Capture 6.1, I've just reset the CC2420. Note that all of the CC2420's registers are cleared to 0x0000. If I had some fancy RF test equipment, I could generate an unmodulated or modulated carrier signal from the CC2420EM by clicking on the Start TX test bar at the bottom of this window. The really useful thing about this window is that we can send and receive raw or IEEE 802.15.4 formatted messages between the pair of CC2420EB dev board/CC2420EM radio board combinations that are part of the CC2420DK.

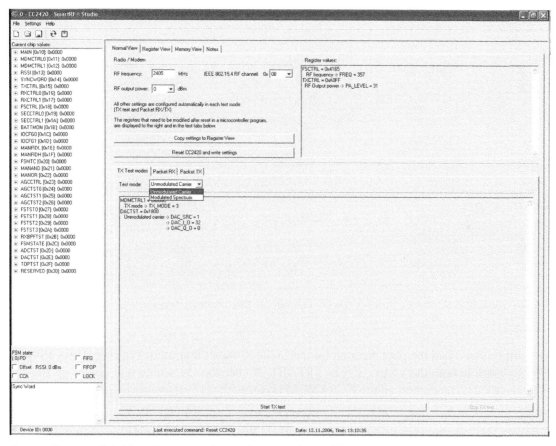

Screen Capture 6.1: There's not much to take away from this SmartRF Studio shot. This screen and its properties are primarily used with RF test equipment to scrutinize the raw RF carrier components generated by the CC2420. We can also use the Packet RX and Packet TX functions to send and receive raw or IEEE 802.15.4 formatted messages.

All of the CC2420 command strobes with the exception of SNOP, which is equivalent to a microcontroller NOP, are represented by buttons across the bottom of Screen Capture 6.2.

As you can see by the unchecked VREG_EN pin checkbox, I haven't powered up the CC2420 yet.

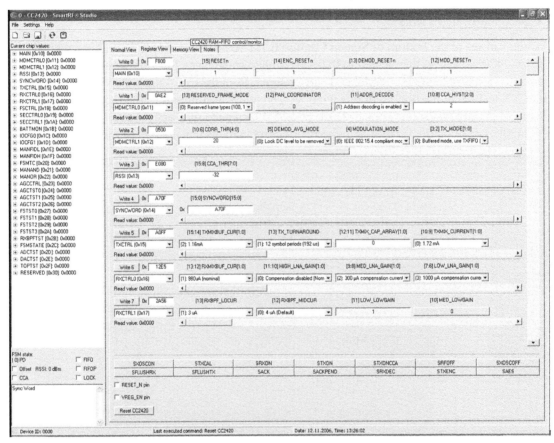

Screen Capture 6.2: The CC2420 is totally inactive at this point as I haven't powered up the IC.

We will follow down the path dictated by the CC2420 state diagram in Figure 6.3. So, I'll put a check mark in both the VREG_EN and RESET_N checkboxes, which will power up and reset the CC2420 mounted on the CC2420EM board.

The view in Screen Capture 6.2 showed the CC2420 register set and allowed us to enter differing register values if we had wished to do so. We're mainly interested in seeing how the CC2420 is used to send and receive IEEE 802.15.4 messages. Diddling with the default register values is not going to get us there at this moment. If register settings were something we were studying, we could have initiated an RX or TX test, which would have set up the registers correctly for us.

After engaging the CC2420's voltage regulator and reset pin, I switched to the SmartRF Studio Memory View in Screen Capture 6.3. The SmartRF Studio Memory View lays out the

CC2420 registers as well as the key CC2420 memory areas such as the TX and RX FIFOs. We can't modify registers from this view, but we can modify the CC2420 memory areas. We can also exercise the CC2420's command strobes from here. The current state of the CC2420 state machine and the current state of the CC2420's CCA and FIFOP pins are also provided in this view.

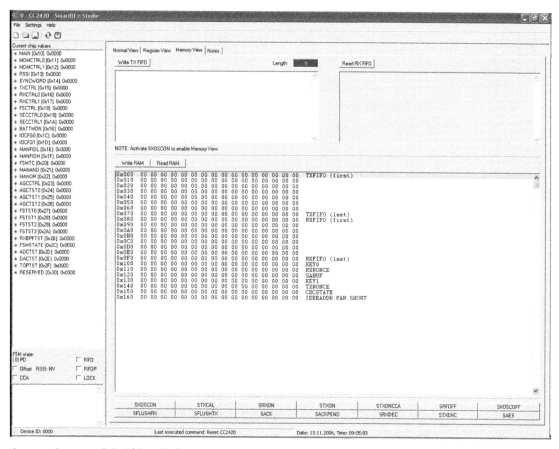

Screen Capture 6.3: This window is a great way to get your arms around what the data looks like inside the CC2420's memory areas. The idea is to allow you to test drive the CC2420 without having to build up a radio or write a CC2420 MAC driver.

Let's throw convention to the wind and just send some bytes out of the CC2420EM's antenna. In Screen Capture 6.4 I typed some ASCII text into the Write TX FIFO window. Clicking on the Write TX FIFO button and attempting to see the bytes show up in the TXFIFO memory area yielded nothing.

According to the state-machine flow chart, the CC2420 is currently in Power Down mode. In Power Down mode the CC2420's crystal oscillator is disabled and access to the CC2420's FIFO and RAM is disabled. It looks like we need to move to IDLE mode and the only way to get there is to issue the SXOSCON command strobe. A click on the SXOSCON button changes the CC2420's state to IDLE and opens up the FIFO and RAM area of the CC2420 in Screen Capture 6.5.

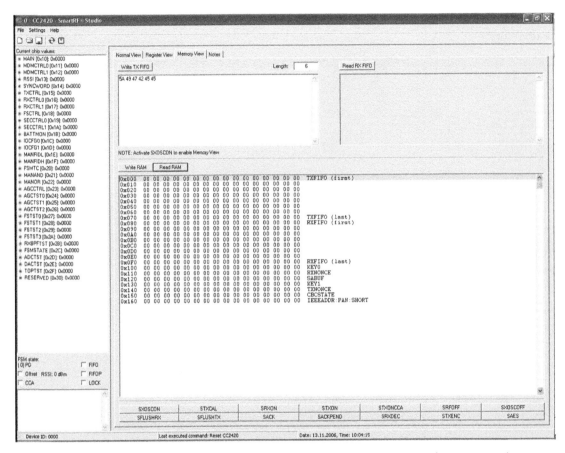

Screen Capture 6.4: After entering the string of ASCII characters, I expected to see my characters inside the TXFIFO memory area. Hmmmm…

The TXFIFO's first byte is the length of my unofficial message, which itself is six bytes in length. The extra two bytes are the CRC bytes, which were inserted automatically by the SmartRF Studio/CC2420 combo. Now that my little unofficial message is lodged inside of the CC2420's TX FIFO, I should be able to do something with it.

According to the CC2420 state machine flow chart, I can issue an STXON command strobe to transmit the contents of the TX FIFO. Before I click on the STXON command strobe button, I'll prepare the Daintree Networks SNA to capture whatever spews out of the CC2420's antenna. Following my click on the STXON command strobe button, the Daintree Networks SNA captured the IEEE 802.15.4 (if you want to call it that) frame shown in Sniffer Capture 6.1.

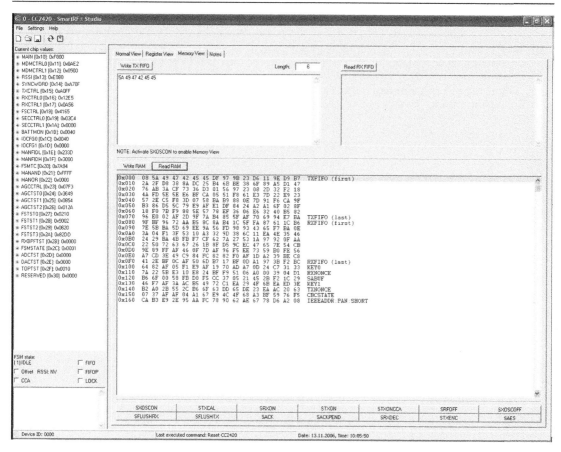

Screen Capture 6.5: Everybody's awake now! After turning on the CC2420's oscillator, I clicked on the Write TX FIFO button one more time and my ASCII string is now part of the TXFIFO.

Sniffer Capture 6.1.

****

```
Frame 1 (Length = 8 bytes)
        Time Stamp: 10:07:09.000
        Frame Length: 8 bytes
        Capture Length: 8 bytes
        Link Quality Indication: 136
IEEE 802.15.4
        Frame Control: 0x495a
                .... .... .... .010 = Frame Type: Acknowledgment (0x0002)
                .... .... .... 1... = Security Enabled: Enabled
                .... .... ...1 .... = Frame Pending: More data
                .... .... ..0. .... = Acknowledgment Request: Acknowledgment
not required
                .... .... .1.. .... = Intra PAN: Within the PAN
                .... ..01 0... .... = Reserved
                .... 10.. .... .... = Destination Addressing Mode: Address
field contains a 16-bit short address (0x0002)
```

```
              ..00 .... .... ....  = Reserved
              01.. .... .... ....  = Source Addressing Mode: Reserved
(0x0001)
     Sequence Number: 71
     Destination PAN Identifier: 0x4542
     Destination Address: 0xff45
     Frame Check Sequence: Correct


0000:   5a 49 47 42 45 45 .. ..                          ZIGBEE..
********************************************************************************
```

Sniffer Capture 6.1: How about that! Ignore everything in the IEEE 802.15.4 Frame Control area and beyond, as this is a bootleg IEEE 802.15.4 message.

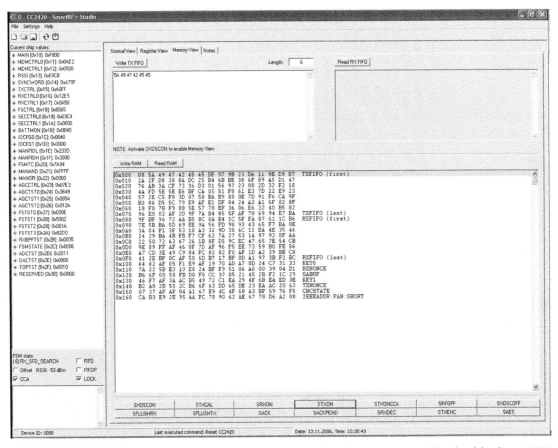

Screen Capture 6.6: Note also that the CCA and LOCK checkboxes are positive in this shot LOCK indicates that the CC2420's PLL has locked. If you still don't know what the CCA box represents, sell this book to your sister.

Even though we succeeded in sending a renegade message from our CC2420EB/CC2420EM evaluation board/radio combination, think about what would happen next in a real IEEE 802.15.4 or ZigBee scenario. Normally, the sending node would want an acknowledgment

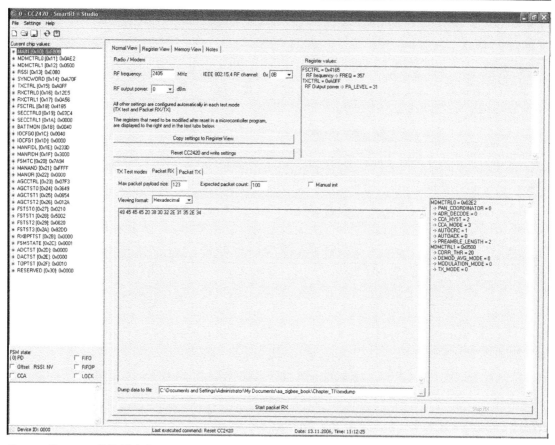

Screen Capture 6.7: This is an easy way to send IEEE 802.15.4 and "unofficial IEEE 802.15.4" data between the pair of nodes that make up the CC2420DK development kit. Note that I dumped the received data to a file in this session. The CC2420 register settings for receive mode are shown on the far right of the shot.

returned from the receiving node. Thus, after a required number of symbol periods the transmitting node would expect to receive an acknowledgment from the receiving node. That means the transmitting node would go into receive mode and await the acknowledgment. Take another look at the CC2420 state-machine diagram in Figure 6.3 and you'll see that the CC2420's state flows from transmit mode to receive mode after the completion of a transmission. This change of states is reflected in the lower left corner of Screen Capture 6.6 as the CC2420's FSM State is listed as [6]RX_SFD_SEARCH, which directly correlates to the CC2420 state-machine flow chart in Figure 6.3.

As you would logically conclude, we can circumnavigate the CC2420's states by simply issuing the appropriate command strobes from the SmartRF Studio windows. The beauty of SmartRF Studio is that you can assemble any type of message that you want and send it along. With the assistance of Sniffer programs like Daintree Networks SNA, you can see what the receiver would get and determine if the IEEE 802.15.4 frame you're sending is bogus

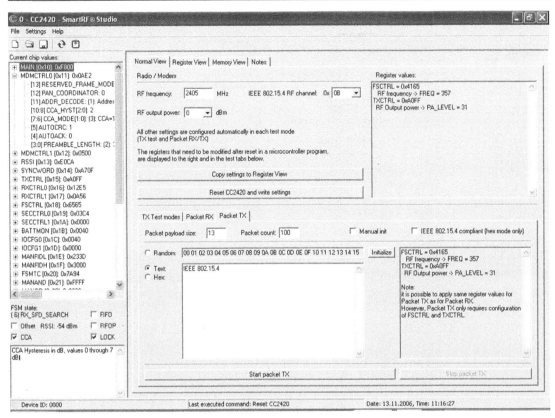

Screen Capture 6.8: Here's a look at the SmartRF Studio transmit session. Everything done here cascades down into the deeper views. For instance, you can go to the memory view and look at the contents of the TXFIFO in raw hex format. Again, note the CC2420 register assistance offered by SmartRF Studio in the far right of this shot.

or IEEE 802.15.4 compliant. SmartRF Studio also provides the basic CC2420 register settings for the carrier, transmit and receive modes in a window next to the test area window. For instance, Screen Capture 6.7 represents a receive test session that I set up on one of the CC2420EB/CC2420EM development platforms, while Screen Shot 6.8 represents the transmit side of the CC2420 equation.

If you're feeling froggy, go on out there and buy yourself one of those pointy moon and star-studded hats the RF guys wear. You've been exposed to the innards of the CC2420 and I hope that you have concluded that if you can set up an SPI interface on your favorite microcontroller, you can communicate with IEEE 802.15.4 using a CC2420. Keep in mind that ZigBee is really a means of providing compatibility at the application levels. There are many ZigBee stack implementations out there that may or may not work in a similar manner. However, every IEEE 802.15.4 implementation that supports a ZigBee stack must work in the same manner. A ZigBee stack exploits the commonality of every IEEE 802.15.4 network. That's why I'm pounding you with more IEEE 802.15.4 stuff than ZigBee stuff right now. If you understand what is going on in the IEEE 802.15.4 part of a ZigBee implementation, you'll be more able to understand why things are the way they are in a ZigBee stack.

Two of Two

What could be better than a CC2420 in an IEEE 802.15.4 or ZigBee network? That's easy. A CC2430. Take everything you've learned about the CC2420 and add an 8051 microcontroller into the mix. The spy satellite was floating overhead of the CC2430 module you see in Photo 6.4.

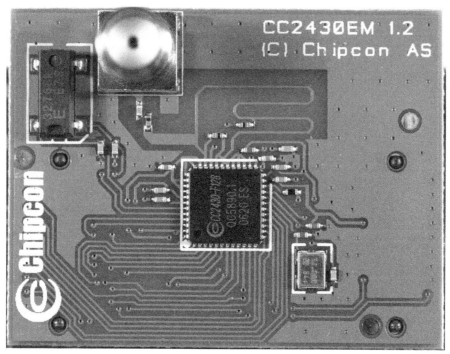

Photo 6.4: No discrete balun?? What gives?

Matching up what you see in Photo 6.4 with the graphic in Figure 6.4 reveals that that little loopy antenna thingie is a folded dipole.

Since there's a microcontroller embedded in the mix, the CC2430's development mother-board is a bit different as well. As you can see in Photo 6.5, the FPGA is gone.

I'm not going to repeat what we did with the CC2420 using the CC2430 as it would look the same. The same development tools I used to make the CC2420 chirp will make the CC2430 chirp as well.

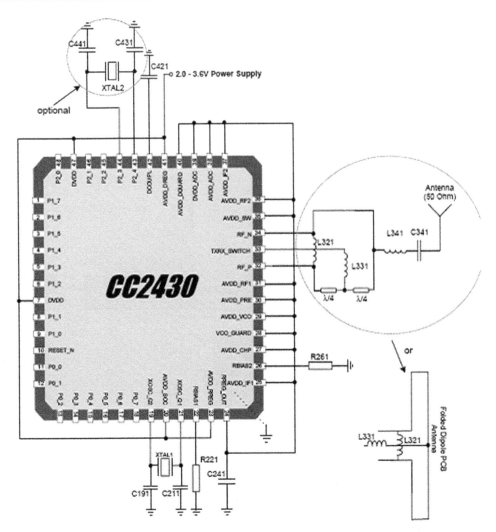

Figure 6.4: This looks absolutely too easy to be RF stuff. I think it says something when the RF components have to be separated from the logic components. Note that all of the RF stuff is to the right of the figure and all of the 8051 stuff is over to the left. That proves my point that RF is evil.

Photo 6.5: Just about anything you want to hook up to a CC2430 is included on this development board. Think about this. All of the smarts for this photo are contained within the CC2430. ALL of them!

About Texas Instruments

It's the Mirrors…I love my DLP television. Thank you, Texas Instruments. Sometimes Texas Instruments milestones:

- The first transistor radio in 1954
- The first integrated circuit circa 1958
- TTL logic circa 1960
- Invented the hand-held calculator 1967
- Invented the single-chip microcomputer 1971
- Invented the single-chip microprocessor 1973

Remember that Texas Instruments calculator that had the magnetic card reader? I had one of those. That rascal cost me about $300 back in the 1970s.

Enough said.

Oh, yes…Arlene. Well, it seems that Arlene was riding around on a date with a young man. She was one year his senior. This wasn't their first date and the young man was behind in the batting count, if you know what I mean. Elvis was the rage and Arlene let her date know it when Elvis's current record was played on the car radio. Wanting to see Arlene again, a young Ricky Nelson turned to Arlene and said, "I'm gonna' make a record, too." Arlene just laughed. The click line and title of one of Ricky's first hits, "I'm Walkin'", is probably what he should have said to Arlene after she shut up.

This one is easy. Where did the Merseybeat sound originate and what band used it to catapult them into history?

Maxstream/XBee

Believe it or not, you've been doing some pretty heavy lifting in the previous chapters. I've thrown all kinds of IEEE 802.15.4 and ZigBee acronyms at you and you've run through MAC and PHY hell with me. We even went caveman and unlocked the key to IEEE 802.15.4 primitives.

Fortunately, you don't have to be a member of the WWE (that's World Wrestling Entertainment for the wrestling-challenged of you out there) or be able to spout specific pieces of the IEEE 802.15.4 and ZigBee specifications on demand to put ZigBee to work for you in a real-world application. Think of it this way. Do you ever really give much thought to what is going on in the bowels of a conversation between your personal computer's USB port and that new Best Buy gadget you just plugged in? How about some serious thoughts about what is going on within that Ethernet cable modem/access point that is enabling your view into the EDTP Electronics web site? Ever really consider pulling your own checksums against that new program you're downloading via FTP? For those of you reading that are ancient enough to remember the reign of the BBS (Bulletin Board System), did you really give a darn about that 300-bps datastream that was being passed between your personal computer's serial port and the modem? NO! Then why should you be excited about the binary exchange within a ZigBee PAN?

You may have your reasons, as it's rather obvious that I do, to delve into the bits and bytes that make up IEEE 802.15.4 and ZigBee conversations. However, in every instance I referenced in the previous paragraph, the ultimate goal is moving the desired data to the next carrier in the communications chain until the data is delivered to the targeted node or endpoint. All you as a user really care about is that what you typed in or uploaded gets to its final destination and if you're on the incoming side of the datastream, you don't want to lose any bytes along the way between your hard drive and the hard drive on the other end of the link. Once your data payload reaches its intended node or endpoint, all you care about is getting a positive response through the same virtual and physical communication paths you pushed the initial request through.

To an engineer, the concepts of USB and Ethernet are easily grasped, as the engineer has been forced to understand the protocols at the bit levels. I'm here to tell you that you really don't have to be a full-blown electrical or software engineer to get your hands around USB, Ethernet, IEEE 802.15.4 or ZigBee. You obviously feel that way too because, engineer or not, you're reading this book, and when you're finished you'll be able to do more with IEEE

802.15.4 networks than just talk about them at the cocktail party. Whether you're a highly skilled engineer or just a guy or gal that wishes to do something with your protocol of interest, try explaining USB enumeration to someone that only knows that you plug in the USB device and it automatically comes online. Or, better yet, tell that same nontechnical end user about all of the wonderful technical stuff going on inside a TCP/IP transaction and how that all relates to that Ethernet cable he or she just plugged into their DSL box. Let's get even closer to home: try to describe IEEE 802.15.4 primitives to a guy or gal project manager that is only interested in getting those temperature readings back from the potato-chip cook tank.

The bottom line is that, in the end, all anyone wants is to get their data and receive their data without having to understand the nuances of every software algorithm and hardware device the data had to traverse. That's the idea behind the MaxStream's XBee and XBee-Pro ZigBee modules.

The XBee ZigBee Module

ZigBee's intended mission is to cut the traditional wires between sensors, traditional wired slave devices and the microcontrollers and microprocessors they serve. Thus, if ZigBee is to emulate a wire, what goes in must come out without any significant change. Unlike the SPI-reliant IEEE 802.15.4-compliant transceivers we've covered thus far, the XBee ZigBee modules employ a UART interface, which allows any microcontroller or microprocessor to immediately use the services of the ZigBee protocol. All the ZigBee hardware designer has to do in this case is make sure that the host's serial-port logic-levels are compatible with the XBee's 2.8–3.4V logic levels. The logic-level conversion can be performed using either a standard RS-232 IC or with logic level translators such as the 74LVTH125 or 74HC125 when the host is directly connected to the XBee UART. Note that I didn't require the host microcontroller to have an on-chip UART. That's because it is a simple thing to emulate a basic UART with firmware. In fact, the Custom Computer Services C compiler has built-in UART emulation facilities aimed at Microchip's PIC microcontrollers. A typical XBee communications link is depicted in Figure 7.1.

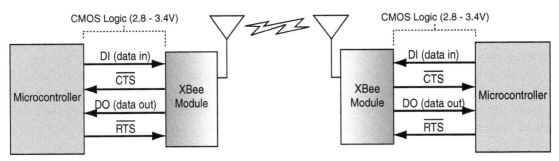

Figure 7.1: Typical XBee communications link. Since ZigBee networks operate at low speeds, the use of a standard RS-232 serial port at each end of a ZigBee or IEEE 802.15.4 network communications link is well within reason.

Data is presented to the XBee module through its DIN pin and must be in the asynchronous serial format, which consists of a start bit, 8 data bits and a stop bit. The XBee modules require the incoming serial signal to idle at a logic high state. Since the input data is going directly into the input of a UART within the XBee module, no RS-232 bit inversions are necessary within the asynchronous serial data stream. If you had a chance to check out my first book, *Networking and Internetworking with Microcontrollers*, you recall that when an asynchronous digital signal is converted to RS-232 signal levels, a "1" is logically high on the digital side and a negative voltage of at least –3V on the RS-232 side. The negative swing can range from –3V to –15V according to the RS-232 specification. The negative state of the RS-232 signal is called a MARK. Conversely, a logical "0" on the digital side is low and termed a SPACE (positive voltage +3V to +15V) on the RS-232 interface. All of the required timing and parity checking is automatically taken care of by the XBee's UART.

As you would expect, the XBee module produces a received data asynchronous serial data stream for the host on its DOUT pin. So, all you need is a simple 3-wire (DIN, DOUT, Ground) serial connection to put ZigBee to work with the XBee and XBee-Pro modules. Just in case you are producing data faster than the XBee can process and transmit it, both XBee modules incorporate a CTS (Clear To Send) function to throttle the data being presented to the XBee module's DIN pin. You can eliminate the need for the CTS signal by sending small data packets at slower baud rates. If you're using the XBee modules in a true ZigBee fashion, the slower speeds and small frames should be automatic.

A simplified view of the XBee internals is represented in Figure 7.2. Incoming data flowing through the DIN pin is buffered by the DIN Buffer until it can be transmitted. You as the programmer and commander have the option to send characters as they enter the DIN pin or buffer up a number of characters to send as a packet. When the XBee module is not sending characters, it can rest in idle mode, enter receive mode, process a command or just sleep it off. The default mode of operation is called Transparent Mode. In Transparent Mode the XBee modules simply act as a serial line replacement. All data passing through the DI pin is queued up for RF transmission and all incoming RF data is piped out of the XBee's DO pin.

Data is automatically buffered in the DI buffer unless the Packetization Timeout (RO) is set to 0 (zero). If RO is set to 0 (zero), data coming into the DI pin is immediately packetized. Otherwise, the incoming data will be buffered in the DI buffer and remain there until:

- No serial characters are received within the RO time-out period.

- The maximum number of characters that will fit into an RF packet (100) is buffered.

- A Command Mode Sequence of (GT + CC + GT) is received.

Hardware flow control for the DI buffer is implemented using the XBee's CTS pin. When the DI buffer is 17 bytes from being full, the CTS line will go logically high to signal the host to cease sending data. When the receiving microcontroller has pulled enough data from the DI buffer to clear 34 bytes, the CTS line is returned to a logical low state, indicating to the transmitting node that the DI buffer can accept more data.

Incoming RF data is placed in the DO buffer before being pushed out of the XBee's DO pin. Hardware flow control for the DO buffer is handled by the RTS pin. Data will not flow from the DO buffer to the host when RTS is logically high.

The bulk of the subsystem boxes shown in Figure 7.2 are contained within two physical ICs, a Freescale Semiconductor MC9S08GT60 microcontroller and a Freescale Semiconductor MC13193 802.15.4 RF transceiver IC.

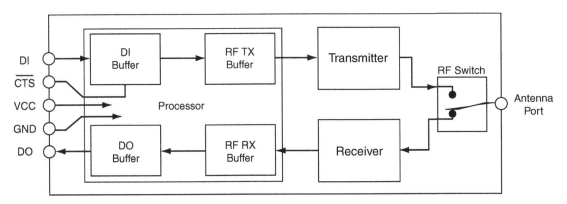

Figure 7.2: Hey! You've run through MAC and PHY hell. You can easily handle this. XBee modules are designed to run right out of the box.

I popped the hood on one of the XBee ZigBee modules to put some meaning behind the boxes in Figure 7.2. The Freescale Semiconductor M9S08GT60 microcontroller and the Freescale Semiconductor MC13193 RF transceiver parts are pretty obvious in Photo 7.1, as they are both clearly marked with the B2 Bomber-like Freescale logo.

Photo 7.1: If you had a chance to read my other book, Implementing 802.11 with Microcontrollers, you will recall that I used a can opener on some 802.11b CompactFlash Wi-Fi cards just as I have here with the XBee module. RF is black magic and I still contend that if you have to shield something, there's something amiss going on behind the metal.

The XBee module's 16-MHz clock source is nestled into the bottom right corner directly under the Freescale MC13193. The MaxStream and Freescale engineers reading this are rolling on the floor laughing at us. Here we are performing SWAGs (call me directly if you don't know what a SWAG is) against obscure electronic components in a picture. They're having fun on us as they really know what the rest of the parts are. Since I only have my intuition to guide me and I don't belong to the pointy-hat-with-moons-and-stars club, I'll put forth my best RF guess that the pair of 6-pin devices just above the Freescale Semiconductor MC13193 are transmit and receive matching transformers. In that there is only one antenna, I'd venture to say that the little 6-pin device just above the Freescale Semiconductor M9S08GT60 microcontroller is an antenna switch, as it is the last active component between the antenna connector and the rest of the RF circuitry. I happen to know that the Freescale Semiconductor MC13193 is really a Freescale Semiconductor MC13192 with an integral ZigBee protocol stack. I also happen to know that the Freescale Semiconductor MC13193 contains receive and transmit buffers to allow the use of smaller microcontrollers like the Freescale Semiconductor M9S08GT60. However, that doesn't exclude the possibility that the Freescale Semiconductor M9S08GT60 microcontroller is also holding a pair of transmit and receive buffers of its own.

The XBee-Pro is basically the same integrated Freescale M9S8GT60/Freescale MC13193 ZigBee radio as the XBee with some extras. I used an electronic can opener (a soldering iron) to pop the hood on the XBee-Pro you see in Photo 7.2.

Photo 7.2: The XBee-Pro is a standard XBee module with an additional adjustable RF power amp, an LNA (low noise amplifier) and an analog-to-digital converter voltage reference switch. The XBee-Pro's RF power amp boosts the XBee-Pro's effective output power to 100 mW.

In Photo 7.2 you can see the 6-pin analog-to-digital converter voltage reference switch IC in the bottom left corner just under the Freescale Semiconductor M9S08GT60 microcontroller. The 50K digital potentiometer that is used to adjust the RF power amp output is the 8-pin package just above the Freescale Semiconductor M9S08GT60 microcontroller. Hopefully, the MaxStream and Freescale RF engineers have dried their laughter-filled eyes as I positively IDed the analog-to-digital converter switch and digital pot devices against the unknown device's datasheets using their package markings. At this point, I would bet that the 16-pin IC just above the Freescale Semiconductor MC13193 is the RF PA (power amplifier) and the 4-lead package to its upper right is the LNA (low noise amplifier) that aids the receiver subsystem of the Freescale Semiconductor MC13193.

Now that you know that a standard Freescale microcontroller and an off-the-shelf Freescale Semiconductor 802.15.4 ZigBee radio are the main ingredients of the XBee module, you realize that since a microcontroller is involved, you have the capability to customize the XBee module's operation. MaxStream also came to that realization and provides some direction as to how to roll your own XBee application. If you're not set up to develop with the HC08 family of Freescale Semiconductor microcontrollers, prepare to dig into the piggy bank. Developing your custom XBee application with the Freescale SMAC can be accomplished with the "free" version of the Metrowerks CodeWarrior Development Studio coupled with the MaxStream Bootloader, which is part of the XBee module's factory firmware. If you decide to go with the Freescale 802.15.4 PHY and MAC development method or the Figure 8 ZigBee stack, you'll need to get a serious copy of the Metrowerks CodeWarrior Development Studio. Hints, libraries and auxiliary files needed to adapt the Freescale SMAC, Freescale 802.15.4 PHY and MAC, and Figure 8 ZigBee stack application methods to the MaxStream PHY are available from the MaxStream web site for a download.

No matter which way you go in XBee development, the MaxStream X-CTU software, which is also a free download from the MaxStream site, allows you to use your personal computer to program the XBee modules serially using the XBee DIN, DOUT, RTS and DTR pins. If you're hardcore, the P&E Microsystems, Inc. USB HCS08/HCS12 Multilink programmer/debugger can also be used to program the XBee modules. The caveat of using the P&E programmer is that you'll overwrite the bootloader code. Losing the bootloader code will also erase the factory calibration information, which allows the XBee-Pro ZigBee modules to attain 18 dBm power output. You'll be limited to 12 dBm output power without the factory calibration info.

Transparent Mode operation using XBee modules may not always be the best choice for an application. We've already seen the usefulness of API calls and how they can take the complication out of writing ZigBee and IEEE 802.15.4 application firmware. To that end, XBee modules can also be controlled via API calls. In the XBee environment, the API mechanism is deployed using structured frames.

For instance, Figure 7.3 is a byte layout of an AT Command API type. The AT Command DL is used to query the lower 32 bits of the XBee modules's 64-bit address. In 16-bit addressing mode (DH = 0x00000000), the DL AT Command queries the XBee module's short address. See, all of the running through hell stuff and introductory ZigBee and IEEE 802.15.4 nomenclature stuff that you've experienced in the previous chapters of this book comes in handy, doesn't it? You'll find that once you've mastered the language and logistics of IEEE 802.15.4 and ZigBee, you will be able to travel to all of the differing ZigBee and IEEE 802.15.4 manufacturer's worlds. Let's use some of our knowledge of IEEE 802.15.4 to assemble a two-man PAN using the XBee modules.

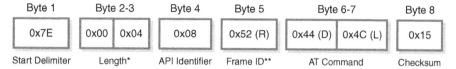

Byte 1	Byte 2-3		Byte 4	Byte 5	Byte 6-7		Byte 8
0x7E	0x00	0x04	0x08	0x52 (R)	0x44 (D)	0x4C (L)	0x15
Start Delimiter	Length*		API Identifier	Frame ID**	AT Command		Checksum

Figure 7.3: This figure is a simple 0x08-Type API, which allows the reading or writing of the target module parameter. In this example, we're simply reading the lower 32 bits of the destination address. The Length is computed from a total of the bytes including the API Identifier up to Checksum. The Frame ID is a randomly selected number.

It always helps to know what you're working with. Photos 7.3 and 7.4 are XBee module interface boards that are provided with the XBee Professional Developer Kit. The XBee USB interface board is shown in Photo 7.3 and the RS-232 version of the XBee interface board can be seen in Photo 7.4.

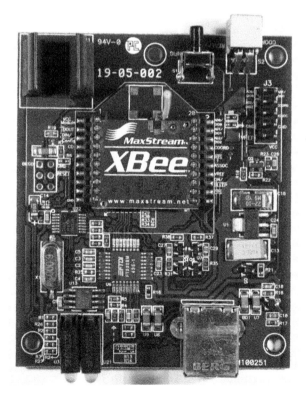

Photo 7.3: Don't get too excited about all of this as there is just enough stuff on this board to power everything and allow the XBee module to communicate with the FTDI RS-232-USB converter IC. Power is supplied via the USB connection.

Photo 7.4: This is the RS-232 version of the XBee interface board. The same idea goes for this board as it does for the USB interface board. There's just enough goo on the board to power the electronics and provide an RS-232 interface to the 9-pin female connector. Note that this board has a power jack.

Both of the XBee interface boards contain nothing fancy and are only designed to make it easy for the XBee programmer to connect the selected XBee module to the outside world.

The XBee interface boards are designed to allow the use of a personal computer in the development of the XBee application. To that end, an XBee personal computer-based control program called X-CTU is provided to eliminate the need to write XBee control routines from scratch. The X-CTU application also provides a debug view of all of the XBee registers and acts as an XBee module programmer. X-CTU even provides a built-in RS-232 terminal emulator.

I found that we can run two instances of X-CTU side by side as you see in Screen Capture 7.1.

As you can see in Screen Shot 7.1, key register values can be changed within the X-CTU windows and written to the XBee modules mounted on the XBee interface boards. I designated the COM3 XBee module as the coordinator and the COM1 XBee module as an end device by setting the XBee module's CE (Coordinator Enable) register to a 1 and 0 (zero), respectively. To keep the pair of IEEE 802.15.4 nodes from jumping around all over the available 2.4-GHz channels, I stuffed a 0x0001 into each of the XBee module's SC (Scan Channels) registers. The single bit in the SC registers only allows channel 11 (0x0B) to be scanned and used. I then assigned a PAN ID of 0xCAFE in the coordinator X-CTU window. The XBee can run in one of three MAC modes:

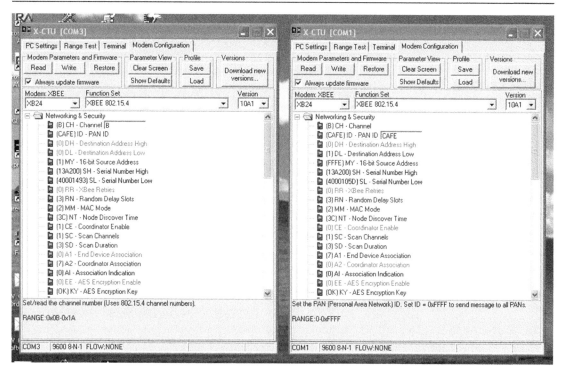

Screen Capture 7.1: The COM3 instance of X-CTU is running against a USB driver that emulates a standard personal computer COM port. In this case, the emulated port is COM3.

- 802.15.4 +MaxStream Header
- 802.15.4 without ACKs
- 802.15.4 with ACKs

I selected 802.25.4 with ACKs for both the coordinator and end device XBee devices by writing a 2 to the XBee's MM (MAC Mode) register. As you well know, association is a key part of assembling a useful PAN. So, on the coordinator side I adjusted the bit pattern within the A2 (Coordinator Association) register to allow association. On the end device side I selected a bit pattern for the A1 (End Device Association) register that instructed the end device XBee module to auto associate.

After writing the changes to both of the potential PAN candidates, I held both of the XBee interface board's reset pins in reset position and released the coordinator's reset pin before releasing the end device's reset pin seconds later. As you have come to expect, I had the Daintree Networks SNA application running the whole time sniffing on channel 11. Following the release of the end device from reset, I watched the packets ripple by in a very familiar pattern in the Daintree Networks SNA Packet List window.

I knew we were on the right track when I saw the coordinator post a Beacon Request to check out the availability of channel 11. The Beacon Request issued by the coordinator is detailed in Sniffer Capture 7.1.

Sniffer Capture 7.1

```
************************************************************************
Frame 206 (Length = 10 bytes)
        Time Stamp: 10:01:56.459
        Frame Length: 10 bytes
        Capture Length: 10 bytes
        Link Quality Indication: 168
IEEE 802.15.4
        Frame Control: 0x0803
                .... .... .... .011  = Frame Type: Command (0x0003)
                .... .... .... 0...  = Security Enabled: Disabled
                .... .... ...0 ....  = Frame Pending: No more data
                .... .... ..0. ....  = Acknowledgment Request: Acknowledgment
not required
                .... .... .0.. ....  = Intra PAN: Not within the PAN
                .... ..00 0... ....  = Reserved
                .... 10.. .... ....  = Destination Addressing Mode: Address
field contains a 16-bit short address (0x0002)
                ..00 .... .... ....  = Reserved
                00.. .... .... ....  = Source Addressing Mode: PAN identifier
and address field are not present (0x0000)
        Sequence Number: 93
        Destination PAN Identifier: 0xffff
        Destination Address: 0xffff
        MAC Payload
                Command Frame Identifier = Beacon Request: (0x07)
        Frame Check Sequence: Correct

0000:    03 08 5d ff ff ff ff 07 .. ..                        ..].......
************************************************************************
```
Sniffer Capture 7.1: Nothing new here. The coordinator is simply blasting away on channel 11 and seeing if someone shoots back.

Once the new PAN was established on channel 11, the end device's active scan of channel 11 led to the release of a Beacon frame from the new PAN Coordinator. If you haven't been chapter hopping, you know that the Beacon frame in Sniffer Capture 7.2 is intended to convey the rules and regulations of the PAN, identify the PAN Coordinator and let the end device know that the PAN has a vacancy.

Sniffer Capture 7.2

```
************************************************************************
Frame 209 (Length = 13 bytes)
        Time Stamp: 10:01:56.745
        Frame Length: 13 bytes
        Capture Length: 13 bytes
```

```
        Link Quality Indication: 232
IEEE 802.15.4
      Frame Control: 0x8000
                    .... .... .... .000  = Frame Type: Beacon (0x0000)
                    .... .... .... 0...  = Security Enabled: Disabled
                    .... .... ...0 ....  = Frame Pending: No more data
                    .... .... ..0. ....  = Acknowledgment Request: Acknowledgment
not required
                    .... .... .0.. ....  = Intra PAN: Not within the PAN
                    .... ..00 0... ....  = Reserved
                    .... 00.. .... ....  = Destination Addressing Mode: PAN
identifier and address field are not present (0x0000)
                    ..00 .... .... ....  = Reserved
                    10.. .... .... ....  = Source Addressing Mode: Address field
contains a 16-bit short address (0x0002)
      Sequence Number: 173
      Source PAN Identifier: 0xcafe
      Source Address: 0x0001
      MAC Payload
            Superframe Specification: 0xcfff
                    .... .... .... 1111  = Beacon Order (0x000f)
                    .... .... 1111 ....  = Superframe Order (0x000f)
                    .... 1111 .... ....  = Final CAP Slot (0x000f)
                    ...0 .... .... ....  = Battery Life Extension: Disabled
                    ..0. .... .... ....  = Reserved
                    .1.. .... .... ....  = PAN Coordinator: Transmitter is
a PAN Coordinator
                    1... .... .... ....  = Association Permit: Coordinator
accepting Association Requests
            GTS Specification: 0x00
                    .... .000  = GTS Descriptor Count (0x00)
                    .000 0...  = Reserved
                    0... ....  = GTS Permit: Coordinator not accepting GTS
Requests
            Pending Address Specification: 0x00
                    .... .000  = Number of short Addresses pending: 0
                    .... 0...  = Reserved
                    .000 ....  = Number of extended Addresses pending: 0
                    0... ....  = Reserved
      Frame Check Sequence: Correct

0000:   00 80 ad fe ca 01 00 ff cf 00 00 .. ..          ..-~J...O....
```
****

Sniffer Capture 7.2: This is what the XBee End Device was looking for. Now there's enough information for the end device to join the PAN.

The XBee End Device has seen enough and writes the check to join the PAN in Sniffer Capture 7.3.

Sniffer Capture 7.3

****

```
Frame 210 (Length = 21 bytes)
        Time Stamp: 10:01:57.058
        Frame Length: 21 bytes
        Capture Length: 21 bytes
        Link Quality Indication: 168
IEEE 802.15.4
        Frame Control: 0xc823
                    .... .... .... .011  = Frame Type: Command (0x0003)
                    .... .... .... 0...  = Security Enabled: Disabled
                    .... .... ...0 ....  = Frame Pending: No more data
                    .... .... ..1. ....  = Acknowledgment Request: Acknowledgment
required
                    .... .... .0.. ....  = Intra PAN: Not within the PAN
                    .... ..00 0... ....  = Reserved
                    .... 10.. .... ....  = Destination Addressing Mode: Address
field contains a 16-bit short address (0x0002)
                    ..00 .... .... ....  = Reserved
                    11.. .... .... ....  = Source Addressing Mode: Address field
contains a 64-bit extended address (0x0003)
        Sequence Number: 96
        Destination PAN Identifier: 0xcafe
        Destination Address: 0x0001
        Source PAN Identifier: 0xffff
        Source Address: 0x0013a2004000105d
        MAC Payload
                Command Frame Identifier = Association Request: (0x01)
                Capability Information: 0x02
                        .... ...0  = Alternate PAN Coordinator: Not capable of
becoming PAN Coordinator
                        .... ..1.  = Device Type: FFD
                        .... .0..  = Power Source: Not receiving power from
alternating current mains
                        .... 0...  = Receiver on when idle: Disables receiver
when idle
                        ..00 ....  = Reserved
                        .0.. ....  = Security Capability: Not capable of using
security suite
                        0... ....  = Allocate Address: Coordinator should not
allocate short address
        Frame Check Sequence: Correct

0000:   23 c8 60 fe ca 01 00 ff ff 5d 10 00 40 00 a2 13    #H`~J....]..@.".
0010:   00 01 02 .. ..                                     .....
```

****

Sniffer Capture 7.3: You already know the drill. This association request provides the end device's vital information to the coordinator whose PAN it wishes to join.

Following the PAN Coordinator's acknowledgment, the end device issues an "empty" data request command in Sniffer Capture 7.4, which is followed by yet another acknowledgment from the PAN Coordinator.

Sniffer Capture 7.4

```
************************************************************************
Frame 212 (Length = 18 bytes)
        Time Stamp: 10:01:57.553
        Frame Length: 18 bytes
        Capture Length: 18 bytes
        Link Quality Indication: 172
IEEE 802.15.4
        Frame Control: 0xc863
                .... .... .... .011 = Frame Type: Command (0x0003)
                .... .... .... 0... = Security Enabled: Disabled
                .... .... ...0 .... = Frame Pending: No more data
                .... .... ..1. .... = Acknowledgment Request: Acknowledgment
required
                .... .... .1.. .... = Intra PAN: Within the PAN
                .... ..00 0... .... = Reserved
                .... 10.. .... .... = Destination Addressing Mode: Address
field contains a 16-bit short address (0x0002)
                ..00 .... .... .... = Reserved
                11.. .... .... .... = Source Addressing Mode: Address field
contains a 64-bit extended address (0x0003)
        Sequence Number: 97
        Destination PAN Identifier: 0xcafe
        Destination Address: 0x0001
        Source Address: 0x0013a2004000105d
        MAC Payload
                Command Frame Identifier = Data Request: (0x04)
        Frame Check Sequence: Correct

0000:   63 c8 61 fe ca 01 00 5d 10 00 40 00 a2 13 00 04    cHa~J..]..@."...
0010:   .. ..                                                            ..
************************************************************************
```
Sniffer Capture 7.4: By "empty" I mean that the Frame Pending bit is indicating no more data.

The acknowledgment to Sequence Number 97 in Sniffer Capture 7.5 informs the end device that there is indeed more data.

Sniffer Capture 7.5

```
************************************************************************
Frame 213 (Length = 5 bytes)
        Time Stamp: 10:01:57.554
        Frame Length: 5 bytes
        Capture Length: 5 bytes
        Link Quality Indication: 232
IEEE 802.15.4
        Frame Control: 0x0012
```

```
                        .... .... .... .010  = Frame Type: Acknowledgment (0x0002)
                        .... .... .... 0...  = Security Enabled: Disabled
                        .... .... ...1 ....  = Frame Pending: More data
                        .... .... ..0. ....  = Acknowledgment Request: Acknowledgment
not required
                        .... .... .0.. ....  = Intra PAN: Not within the PAN
                        .... ..00 0... ....  = Reserved
                        .... 00.. .... ....  = Destination Addressing Mode: PAN
identifier and address field are not present (0x0000)
                        ..00 .... .... ....  = Reserved
                        00.. .... .... ....  = Source Addressing Mode: PAN identifier
and address field are not present (0x0000)
        Sequence Number: 97
        Frame Check Sequence: Correct


0000:   12 00 61 .. ..                                            ..a..
```

****
Sniffer Capture 7.5: Not so fast, end device. I'm still working on this association request you submitted.

Seems that the XBee End Device's check cleared and the PAN Coordinator hands over a key to a room in the PAN in Sniffer Capture 7.6.

Sniffer Capture 7.6

****
```
Frame 214 (Length = 27 bytes)
        Time Stamp: 10:01:57.557
        Frame Length: 27 bytes
        Capture Length: 27 bytes
        Link Quality Indication: 232
IEEE 802.15.4
        Frame Control: 0xcc63
                        .... .... .... .011  = Frame Type: Command (0x0003)
                        .... .... .... 0...  = Security Enabled: Disabled
                        .... .... ...0 ....  = Frame Pending: No more data
                        .... .... ..1. ....  = Acknowledgment Request: Acknowledgment
required
                        .... .... .1.. ....  = Intra PAN: Within the PAN
                        .... ..00 0... ....  = Reserved
                        .... 11.. .... ....  = Destination Addressing Mode: Address
field contains a 64-bit extended address (0x0003)
                        ..00 .... .... ....  = Reserved
                        11.. .... .... ....  = Source Addressing Mode: Address field
contains a 64-bit extended address (0x0003)
        Sequence Number: 83
        Destination PAN Identifier: 0xcafe
        Destination Address: 0x0013a2004000105d
        Source Address: 0x0013a20040001493
        MAC Payload
                Command Frame Identifier = Association Response: (0x02)
                Short Address: 0xfffe
```

```
            Association Status: Association Successful (0x00)
        Frame Check Sequence: Correct

0000:    63 cc 53 fe ca 5d 10 00 40 00 a2 13 00 93 14 00    cLS~J]..@.".....
0010:    40 00 a2 13 00 02 fe ff 00    .. ..
@.".....~....
```
****
Sniffer Capture 7.6: Everybody's happy here. Think about this. Most of the bytes that make up this frame are IEEE address bytes.

I often wonder if the Hayes modem folks ever in their wildest dreams expected to see their "AT" command interface tacked onto ZigBee radios. Well, if you don't have a need to roll your own XBee embedded application using a full-blown ZigBee stack such as Figure 8, the venerable XBee "AT" command set can be used to modify and read the XBee module parameters. The XBee's "AT" command set and its use of the RS-232 protocol make the XBee and XBee-Pro prime candidates for use with a UART-equipped microcontroller. So, I threw together a little microcontroller-based board complete with an RS-232 serial port to collect the data payloads from the IEEE 802.15.4 packets flying between the XBee modules.

The overall long-distance ZigBee/IEEE 802.15.4 scheme I concocted is simple. An XBee node is either running an embedded application or is under control of a host microcontroller, which is also most likely running an embedded application. Either way, data is passed from the remote XBee node to a central collection XBee node, which passes the received ZigBee/IEEE 802.15.4 node data to the microcontroller on the printed circuit board you see in Photo 7.5. The receiving microcontroller can then process the data or just pass it along to the Ethernet engine encapsulated within a UDP datagram. By adding Ethernet and UDP, small data payloads can now flow from PAN to LAN to WAN and vice versa.

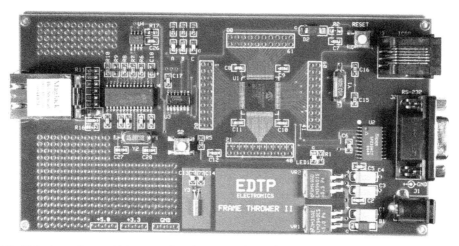

Photo 7.5: This is really an Ethernet development board. However, this little piece of electronic wizardry contains a realtime clock, a serial port, a dual-voltage logic power supply and a debugging/programming port for its PIC18LF8722 microcontroller. The idea is to pipe serial data in from the XBee module and push it out of the Ethernet development board's Ethernet port.

Let's examine what it takes to perform the PAN-to-LAN data transfers. To keep things easy, I put the XBee modules into Transparent Mode in which the serial datastream that enters the XBee module DIN input is not held in a buffer and is transmitted as quickly as possible. The microcontroller on my ZigBee/IEEE 802.15.4-to-Ethernet bridge is programmed to catch every incoming serial character by way of interrupt and stuff the captured characters into a receive buffer. The buffered characters are then punched into the data area of a UDP datagram, which is preconstructed and lying in wait in the microcontroller's SRA7. The details of the construction of the UDP datagram and its ultimate transmission are laid out in Listing 7.1. If you're not familiar with UDP, I've provided a UDP Primer on the Companion website just for you.

Listing 7.1

```
//****************************************************************
//*   This function sends UDP packet with user defined payload
//****************************************************************
void send_udp_datagram()
{
    char temp;
    unsigned int buffer_addr,i,tx_end,tx_cnt;

    tx_end = TXSTART;
    //load beginning page for transmit buffer
     banksel(EWRPTL);
    wr_reg(EWRPTL,LOW_BYTE(TXSTART));
    wr_reg(EWRPTH,HIGH_BYTE(TXSTART));

    //write ENC28J60 control byte
    wr_sram(tx_control_byte);
    ++tx_end;
    //build destination MAC address
    udp_packet[enetpacketDest0]=remotemacaddrc[0];
    udp_packet[enetpacketDest1]=remotemacaddrc[1];
    udp_packet[enetpacketDest2]=remotemacaddrc[2];
    udp_packet[enetpacketDest3]=remotemacaddrc[3];
    udp_packet[enetpacketDest4]=remotemacaddrc[4];
    udp_packet[enetpacketDest5]=remotemacaddrc[5];
    tx_end += 6;
    //build source MAC address
    udp_packet[enetpacketSrc0]=macaddrc[0];
    udp_packet[enetpacketSrc1]=macaddrc[1];
    udp_packet[enetpacketSrc2]=macaddrc[2];
    udp_packet[enetpacketSrc3]=macaddrc[3];
    udp_packet[enetpacketSrc4]=macaddrc[4];
    udp_packet[enetpacketSrc5]=macaddrc[5];
    tx_end += 6;
    //build packet type (IP)
    udp_packet[enetpacketType0] = 0x08;
    udp_packet[enetpacketType1] = 0x00;
    tx_end += 2;
    //build version, length,tos,packet length
```

```
udp_packet[ip_vers_len] = 0x45;
udp_packet[ip_tos] = 0x00;
udp_packet[ip_pktlen] = 0x00;
udp_packet[ip_pktlen+1] = 0x24;
tx_end += 4;
//build packet ID
++cntri;
udp_packet[ip_id] = make8(cntri,0);
udp_packet[ip_id+1] = make8(cntri,1);
tx_end += 2;
//build fragment offset
udp_packet[ip_frag_offset] = 0x00;
udp_packet[ip_frag_offset+1] = 0x00;
tx_end += 2;
//build time to live, UDP protocol ID
udp_packet[ip_ttl] = 0x80;
udp_packet[ip_proto] = 0x11;
//build IP source and destination addresses
for(i=0;i<4;++i)
{
    udp_packet[ip_srcaddr+i]=ipaddrc[i];
    udp_packet[ip_destaddr+i]=remoteipaddrc[i];
    tx_end += 2;
}
//calculate the IP header checksum
udp_packet[ip_hdr_cksum]=0x00;
udp_packet[ip_hdr_cksum+1]=0x00;
hdr_chksum =0;
hdrlen = (udp_packet[ip_vers_len] & 0x0F) * 4;
addr = &udp_packet[ip_vers_len];
cksum();
chksum16= ~(hdr_chksum + ((hdr_chksum & 0xFFFF0000) >> 16));
udp_packet[ip_hdr_cksum] = make8(chksum16,1);
udp_packet[ip_hdr_cksum+1] = make8(chksum16,0);
tx_end += 4;
//build UDP source port address
udp_packet[UDP_srcport] = 0x13;
udp_packet[UDP_srcport+1] = 0x88;
tx_end += 2;
//build UDP destination port address
udp_packet[UDP_destport] = 0x00;
udp_packet[UDP_destport+1] = 0x07;
tx_end += 2;
//build UDP packet length
udp_packet[UDP_len] = 0x00;
udp_packet[UDP_len+1] = 0x10;
tx_end += 2;
//udp_packet[UDP_cksum] goes here
tx_end += 2;
//UDP data goes here
 for(i=0;i<8;++i)
 udp_packet[UDP_data+i] = 0x30 + i;
```

```
    tx_end += 8;
      //calculate the UDP checksum
      udp_packet[UDP_cksum] = 0x00;
      udp_packet[UDP_cksum+1] = 0x00;
     hdr_chksum =0;
     hdrlen = 0x08;
     addr = &udp_packet[ip_srcaddr];
     cksum();
     hdr_chksum = hdr_chksum + udp_packet[ip_proto];
     hdrlen = 0x02;
     addr = &udp_packet[UDP_len];
     cksum();
     hdrlen = make16(udp_packet[UDP_len],udp_packet[UDP_len+1]);
     addr = &udp_packet[UDP_srcport];
     cksum();
     chksum16= ~(hdr_chksum + ((hdr_chksum & 0xFFFF0000) >> 16));
     udp_packet[UDP_cksum] = make8(chksum16,1);
     udp_packet[UDP_cksum+1] = make8(chksum16,0);
    //mark end of UDP datagram
    banksel(ETXNDL);
    wr_reg(ETXNDL,LOW_BYTE(tx_end));
    wr_reg(ETXNDH,HIGH_BYTE(tx_end));
    tx_cnt = tx_end - TXSTART;
    //write UDP datagram to ENC28J60 Transmit Buffer
    for(buffer_addr=0;buffer_addr<tx_cnt+1;++buffer_addr)
        wr_sram(udp_packet[enetpacketDest0 + buffer_addr]);
    //send the contents of the transmit buffer onto the network
    if(1 == (rd_reg8(EIR) & EIR_TXERIF))
    {
     bfc_reg(EIR, EIR_TXERIF);
      bfs_reg(ECON1, ECON1_TXRTS);
     bfc_reg(ECON1, ECON1_TXRTS);
    }
    bfc_reg(EIR,EIR_TXIF);
    bfs_reg(ECON1, ECON1_TXRTS);
    while(1 == (rd_reg8(ECON1) & ECON1_TXRTS));
    ++tx_end;

    banksel(ERDPTL);
    wr_reg(ERDPTL,LOW_BYTE(tx_end));
    wr_reg(ERDPTH,HIGH_BYTE(tx_end));
     for(i=0;i<7;++i)
        tx_status[i]= rd_sram();
    //temp = rd_reg8(ESTAT);
    if(1 == (rd_reg8(ESTAT) & ESTAT_TXABRT))
        printf("\r\nUDP Datagram Transmission Aborted..");

}
```

Listing 7.1: Don't worry. The complete code listing is on the Companion website. The code is intended to control the functionality of the EDTP Frame Thrower II development board shown in Photo 7.5. If UDP is a foreign word to you, take a look at the UDP Primer on the Companion website.

The new Microchip ENC28J60 is the Ethernet engine that drives the ZigBee/IEEE 802.15.4 bridge board. The ENC28J60's 8K of internal packet-buffer area can be carved up by loading certain extent registers with the buffer extent values. I allocated 1K of transmit-buffer area, which left the remaining 7K for receive-buffer duty. TXSTART represents the beginning of the transmit-buffer area within the ENC28J60. Rather than tie up ENC28J60 buffer space, I assembled a UDP datagram within the microcontroller memory with the send_udp_datagram function shown in Listing 7.1. A UDP datagram is a very simple collection of destination/ source hardware and IP addresses, checksums, destination/source logical port addresses and data. I build a complete UDP datagram from top to bottom and send it within the send_udp_ datagram function's code. The data area of the UDP datagram is shown as static values in Listing 7.1.

Before we can send that little UDP datagram and do something useful with the data it carries, we must have a defined destination. So, I've cooked up a little code module to seek out and map the destination Ethernet station. The arp_request function shown in Listing 7.2 writes the contents of an ARP request packet directly into the ENC28J60's transmit buffer area. There is no reason to build the ARP request packet up in microcontroller memory as it is only used once at the beginning of the process.

Listing 7.2

```
//**************************************************************
//* Perform ARP Request
//*    This routine uses a known remote IP address to get a remote
//*    Ethernet modules's MAC (hardware) address.
//**************************************************************
void arp_request(void)
{
    //char temp;
    unsigned int i,tx_end;

    clr_arpflag;
    tx_end = TXSTART;
    //load beginning page for transmit buffer
     banksel(EWRPTL);
    wr_reg(EWRPTL,LOW_BYTE(TXSTART));
    wr_reg(EWRPTH,HIGH_BYTE(TXSTART));

    //write control byte
    wr_sram(tx_control_byte);
    ++tx_end;
    //build destination MAC address
    for(i=0;i<6;++i)
    {
        wr_sram(0xFF);
        ++tx_end;
    }
    //build source MAC address
```

```
for(i=0;i<6;++i)
{
    wr_sram(macaddrc[i]);
    ++tx_end;
}

wr_sram(0x08);        //ARP packet type
wr_sram(0x06);
tx_end +=2;
wr_sram(0x00);        //hardware type = 10Mb Ethernet
wr_sram(0x01);
tx_end +=2;
wr_sram(0x08);        //IP protocol
wr_sram(0x00);
tx_end +=2;
 wr_sram(0x06);       //hardware addr len (06)
 wr_sram(0x04);       //hardware protocol addr len(04)
tx_end +=2;
wr_sram(0x00);        //ARP request
wr_sram(0x01);
tx_end += 2;
//build source MAC address
for(i=0;i<6;++i)
{
    wr_sram(macaddrc[i]);
    ++tx_end;
}
//build source IP address
for(i=0;i<4;++i)
{
    wr_sram(ipaddrc[i]);
    ++tx_end;
}
//build unknown target MAC address area
for(i=0;i<6;++i)
{
    wr_sram(0x00);
    ++tx_end;
}
//build target IP address
for(i=0;i<4;++i)
{
    wr_sram(remoteipaddrc[i]);
    ++tx_end;
}
//mark end of ARP request packet
banksel(ETXNDL);
wr_reg(ETXNDL,LOW_BYTE(tx_end));
wr_reg(ETXNDH,HIGH_BYTE(tx_end));
//send the contents of the transmit buffer onto the network
while((rd_reg8(ECON1) & ECON1_TXRTS) == 1);
```

```
   if((rd_reg8(EIR) & EIR_TXERIF) == 1)
   {
    bfc_reg(EIR, EIR_TXERIF);
     bfs_reg(ECON1, ECON1_TXRTS);
    bfc_reg(ECON1, ECON1_TXRTS);
   }
   //bfc_reg(EIR,EIR_TXIF);
   bfs_reg(ECON1, ECON1_TXRTS);
}
```

✱✱

Listing 7.2: Again, don't worry. If you don't speak ARP, you can get a first hand ARP tutorial from the ARP Primer I've included on the Companion website for you.

The method of updating and sending the ZigBee or IEEE 802.15.4 data inside the UDP datagrams is dependent on your application. I've included real-time clock hardware as well as a real-time clock driver in the ZigBee/IEEE 802.15.4 bridge application code to allow you to send the UDP datagrams at timed intervals. You may also choose to kick off a UDP datagram after collecting a certain amount of data. For demonstration purposes, I coded my ZigBee/IEEE 802.15.4 bridge application skeleton to send a UDP datagram every minute.

I used the null modem adapter that came with the XBee Professional Developer Kit to attach the RS-232-equipped XBee development pod to the serial port of my ZigBee/IEEE 802.15.4 bridge printed circuit board. In lieu of a microcontroller/XBee module lashup at the remote end of the PAN, I used the XBee Development Kit's USB-attached XBee development pod and my laptop to simulate the remote data collecting ZigBee node. The experimental lashup is shown in Photo 7.6.

Photo 7.6: The null modem adapter crosses the RS-232 lines to satisfy the communication port needs of both the Frame Thrower II and the XBee interface board, which are both wired as DCE devices.

Although very powerful when coupled with an external host microcontroller, the XBee and XBee-Pro modules can be used effectively as stand-alone PAN devices, as the Freescale Semiconductor microcontroller at the heart of the XBee modules is capable of doing other microcontroller things such as analog-to-digital conversion and general-purpose I/O. The XBee command set includes directives to manipulate these analog and digital resources.

About MaxStream

MaxStream and its products have been featured in cover stories and special reports of magazines read throughout the world. That's right. I've done a few MaxStream magazine columns myself. Here's another company that listens and responds. During my time writing this chapter, I "killed" one of my MaxStream XBee modules. Well, at least I thought I had killed it. So, I got online and went to the MaxStream site looking for help. After I couldn't figure it out on my own (I know, that's sad isn't it?), I clicked onto their online support line. Within a minute I was chatting with an engineer that walked me through reviving my seemingly dead XBee module. MaxStream was also one of the very first responders with content for this book.

You now have a very good idea of what the XBee radios are all about. Let's move on and take the XBee technology one step further by mating an XBee module with a Rabbit.

Wait, don't you want to know about the Merseybeat sound? Well, it seems that these young men were doing time in Germany. Not that kind of time, musical time. They had progressed from being The Quarrymen, to the Silver Beatles, to just the Beatles by the time they were honing their chops in the red-light district of Hamburg, Germany. Believe it or not, the Beatles were just another Liverpool band before they went to Hamburg. However, when they returned as a really tight band called the Beatles, the folks in the Merseyside area that frequented the Cavern Club listened to their Merseybeat sound night after night.

Here's a guitar question for you. George Harrison was considered "the" guitar player for the Beatles. Which very unlikely guitar did George cherish as a young Beatle?

Hopping Down the Bunny Trail

As many of you know, I've been putting words and pictures into print for quite some time now. I was brought into this business rather unexpectedly when my friend, the late Bill Green of *Radio Electronics* and *Popular Electronics* fame, helped me to get my first real magazine-writing gig some 23 years ago. Long before the email and internet craze began, Bill and I would spend evenings at his place going over the day's editorial happenings. Usually, we would discuss the reader feedback he had received that day and wonder why some of the things readers were thinking got onto a piece of paper and into our hands. I learned quickly that as a technical writer there are a few folks out there spending their reading time in an effort to discredit you. Usually, those guys would send nasty notes to the magazine editor or fire off a really nasty and degrading letter to the author. Bill's technology was sound and he was a student of electronics. As a consequence, the naysayers never won a battle with Bill and moved on to what they thought would be an easier target. Fortunately, I learned from Bill and I do my research as best I can before committing my ideas to print. However, I came upon an internet forum discussion about me that really hurt. One of the forum members was commenting on how I "like everything" and never write negatively about the stuff I evaluate in my columns. He went on to say that "this guy even likes Rabbit microcontrollers." He followed up that statement by disparaging the Rabbit Semiconductor microcontroller and its technology. That's one of the reasons why I don't participate regularly in forums to this day. I sucked it up just as Bill would have and moved on. Despite the forum member's attack on Rabbit Semiconductor and me, Rabbit Semiconductor's business model and microcontrollers are still going strong and I'm still here writing books and magazine columns. As far as "liking everything I write about," I will never waste your time or mine by writing about things that don't work and shouldn't be given the time of day.

Say what you will about Rabbit microcontrollers. However, if you really want to understand how something works, purchase an associated Rabbit Development kit. In my opinion, the most comprehensive 802.11b development kit in the world is sold by Rabbit Semiconductor. If you want to learn to use the internet protocols and post web pages with a microcontroller, get a Rabbit Semiconductor's Ethernet development kit and learn how to do it the right way. My reasons for feeling this way involve the way that the Rabbit code and libraries are structured. Rabbit libraries are not compiled object files, Rabbit libraries are actual source code files that you can open with an editor, read and study. Another Rabbit plus in my book is that every Rabbit development kit I've ever come into contact with works as designed with no compromises. The Rabbit ZigBee/802.15.4 Application Kit is no exception to the rule of the Rabbit.

I purposely put the XBee chapter before this one as the Rabbit ZigBee/802.15.4 Application Kit is built around XBee IEEE 802.15.4 radio technology. I also wanted to expose you to the X-CTU tool and the ways of the XBee modules, so you would be able to immediately transfer that knowledge to the concepts to be offered up in this chapter.

Let's take a look at the Rabbit ZigBee/802.15.4 Application Kit hardware. The XBee RF Module interface board that is standard equipment with the Rabbit ZigBee/802.15.4 Application Kit is shown in Photo 8.1.

Photo 8.1: XBee RF Module interface board. The only thing you don't see is the voltage regulator, which is nestled under the XBee module. Otherwise, there's an RS-232 converter IC, the basic interconnects and some user switches and LEDs.

The Rabbit ZigBee/802.15.4 Application Kit XBee RF Module interface board is very simple in design as it is only intended to provide easy access to the XBee module's serial interface and power. A well-heatsinked voltage regulator resides underneath the XBee module with nothing mounted on the back side of the XBee RF Module interface board. There's just enough on the XBee RF Module interface board to allow the programmer access to the XBee's serial interface and power connections. The serial connector to the far left is connected directly to the XBee serial I/O pins while the RS-232 connector at the bottom of the shot is buffered by the RS-232 converter IC to its left. The Rabbit ZigBee/802.15.4 Application Kit XBee RF Module interface board can accommodate the XBee-Pro radio module as well.

Photo 8.2 is an assemblage of a RabbitCore RCM3720 mounted on a RabbitCore RCM3720 Prototyping Board. The Rabbit ZigBee/802.15.4 Application Kit XBee interface module is simply sitting in for the photo shoot. The Rabbit ZigBee/802.15.4 Application Kit XBee interface module connects to the RabbitCore RCM3720 via the RS-232 connector at the far right of the RabbitCore RCM3720 Prototyping Board. Power for the XBee interface module is stolen from the RabbitCore RCM3720 Prototyping Board power rail. I'm not going to get deep into a discussion of the RabbitCore RCM3720 but I will tell you it is a Rabbit 3000-based microcontroller platform with 512K of Flash and 512K of SRAM. The RabbitCore RCM3720 also includes a fully functional 10-Mbps Ethernet interface, which is based on the RTL8019AS. The RabbitCore RCM3720 offers 33 general-purpose I/O lines and 4 serial ports. Although we have the luxury of mounting our RabbitCore RCM3720 on a RabbitCore RCM3720 Prototyping Board, the RabbitCore RCM3720 is really designed to have its .1-inch 2 × 20 dual-row IDC header plugged onto a user-unique production motherboard.

Photo 8.2: RabbitCore RCM3720 on a RabbitCore RCM3720 Prototyping Board. The RabbitCore RCM3720 is primarily used in Ethernet applications. However, since the XBee is a serial device and the RabbitCore RCM3720 sports 4 serial ports, the RabbitCore RCM3720 can offer some really interesting XBee application options.

The XBee module in Photo 8.2 connects to the RabbitCore RCM3720's serial port by way of ribbon cable. Rabbit firmware is written using Rabbit Semiconductor's Dynamic C compiler, which also integrates the microcontroller debugging environment. The XBee application firmware that runs on the Rabbit microcontroller is based on the XBee AT command set and not the XBee API. The Rabbit ZigBee/802.15.4 Application Kit includes libraries that have taken all of the XBee AT commands and assembled them into simple function calls. The parameters of the XBee AT command set are entered as function call arguments in the Dynamic C source code. Effectively, the XBee AT command set has been converted to function calls that return values solicited by the AT commands. Let's wander through a simple Rabbit ZigBee/802.15.4 Application Kit Dynamic C-based XBee application and see if we can figure out what's going on behind the scenes.

The first order of business is to use X-CTU to configure the XBee module I/O to match the RF Module interface board hardware. The Rabbit ZigBee/802.15.4 Application Kit RF Module interface boards are physically configured as follows:

- DIO0 = Output DS1 LED
- DIO1 = Output DS2 LED
- DIO2 = Input S1 pushbutton switch
- DIO3 = ADC BAT battery voltage monitor
- DIO4 = Input S2 pushbutton switch

Thus, we can use X-CTU to set up the XBee module's I/O pins this way:

- D0 - DIO0 Configuration = 4 (output low)
- D1 - DIO1 Configuration = 4 (output low)
- D2 - DIO2 Configuration = 3 (input)
- D3 - DIO3 Configuration = 2 (ADC)
- D4 - DIO4 Configuration = 3 (input)

Rather than babble along trying to explain the numbers behind the DIOX Configurations, I captured an X-CTU session in Screen Capture 8.1 for you.

Only the XBee End Device gets the general-purpose I/O configuration treatment via the X-CTU. The XBee coordinator is set up using X-CTU to allow devices to join the PAN and the XBee End Device is configured via X-CTU to automatically join the PAN. To make sure things were not out of my control, the PAN ID is preset to 0xAAAA in both the PAN Coordinator and end device. I also shut down every channel except channel 11 by setting the SC (Scan Channels) parameter to 0x0001.

One of the non-IEEE 802.15.4 things that can be configured on an XBee module is a Node ID. In the XBee world a Node ID (NI) is an ASCII name that is associated with the node. In our little peer-to-peer network the PAN Coordinator has an NI of DIO-COORD and I reached into my Parliament-Funkadelic bag and assigned an NI of STARCHILD-1 to the end device. (For those of you asking what the heck is a Parliament-Funkadelic, they are two bands in one formed by George Clinton in the late 1960s. You can hear Star Child speak about his preference for funk music on Parliament's 1975 release "Mothership Connection." I wore out two copies of this album.) OK…Everything is set up hardware-wise and you have a new album to buy.

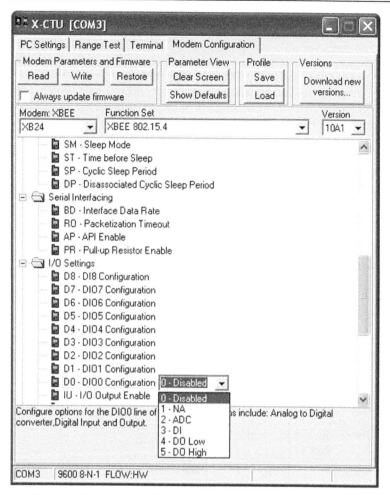

Screen Capture 8.1: Just in case you don't have your Rabbit ZigBee/802.15.4 Application Kit and X-CTU application handy, this shot should clear up any XBee general-purpose I/O configuration questions.

I'll use Dynamic C's STDIO window to show you the results of the execution of the XBee library function calls. The application we will be examining uses the RabbitCore RCM3720's D serial port.

Code Snippet 8.1

```
**************************************************************************
#define ATCMDRSP_SP      D          //set to serial port A, B, C, D, E, or F
#define DINBUFSIZE       255        //PC1 = RxD -- XBee pin 2 = Dout
#define DOUTBUFSIZE      127        //PC0 = TxD -- XBee pin 3 = Din
#define SERD_RTS_PORT    PCDR       //RTS is output flowcontrol
#define SERD_RTS_SHADOW  PCDRShadow
```

```
#define SERD_RTS_BIT     2        //PC2
#define SERD_CTS_PORT    PCDR     //CTS is input flowcontrol
#define SERD_CTS_BIT     3        //PC3
#define DEFAULTBAUD      9600L    //xbee factory default baud rate
```
****

Code Snippet 8.1: Another big plus for Rabbit is that their code is really easy to read and understand. For instance, ATCMDRSP_SP or AT CoMmanD RESponse_Serial Port is assigned to serial port D on Rabbit general-purpose I/O Port C.

If you've never programmed in Dynamic C, you've probably never closely examined the layout of the Rabbit microcontrollers' general-purpose I/O logic. Figure 8.1 is a graphical depiction of the RabbitCore RCM3720's Port C, which is synonymous with PCDR in Code Snippet 8.1.

PIN	DEFAULT	PRIMARY FUNCTION		ALTERNATE FUNCTION		CAPABILITY
		OUTPUT	INPUT	OUTPUT	INPUT	
PC7	←		INPUT		RXA	←
PC6	→	OUTPUT		TXA		→
PC5	←		INPUT		RXB	←
PC4	→	OUTPUT		TXB		→
PC3	←		INPUT		RXC	←
PC2	→	OUTPUT		TXC		→
PC1	←		INPUT		RXD	←
PC0	→	OUTPUT		TXD		→

Figure 8.1: Graphical depiction of RabbitCore RCM3720's Port C. The code in Code Snippet 8.1 is written around the capabilities of the PCD8. Don't worry—if you're Rabbit challenged, I've provided a Rabbit microcontroller system reference poster on the Companion website.

The next step in the chain involves setting the initial baud rate of the Rabbit microcontroller's serial port, enabling the flow control (RTS/CTS) and flushing the serial port buffers. That's all done in Code Snippet 8.2.

Code Snippet 8.2
****
```
brdInit();
serOpen(ATCMDRSP_SP,DEFAULTBAUD);
serFlowCtrlOn(ATCMDRSP_SP);          //enable flow control
serWrFlush(ATCMDRSP_SP);
serRdFlush(ATCMDRSP_SP);
```
****

Code Snippet 8.2: The brdInit function has been present in every other Rabbit development kit I've had experience with. The initial operational states of the Rabbit microcontroller's general-purpose I/O are established within the brdInit function's code. I'm rather sure that the rest of the code in this snippet is self-explanatory.

The next piece of code uses one of the XBee library functions, xb_atModeOn, in an attempt to contact the XBee PAN Coordinator module, which is serially attached to the RabbitCore RCM3720. The idea behind Code Snippet 8.3 is to send the initial "AT <Enter>" and get the "OK" response, which will verify that the correct baud rate is being used. Baud rates of 9600 bps and 115200 bps are attempted. If things blow up during the process, the application will halt in its tracks.

Code Snippet 8.3

★★

```
printf("Trying DEFAULTBAUD (%ld) ",DEFAULTBAUD);
if(xb_atModeOn(1500)<0)                  // if fails try 115200
{  printf("FAILED, trying (115200L) ");
serOpen(ATCMDRSP_SP,115200L);
   serWrFlush(ATCMDRSP_SP);
    serRdFlush(ATCMDRSP_SP);
if(xb_atModeOn(1500)<0)
    {   printf("\nTried 9600 baud and 115200 baud and Failed\n");
       exit(0);
    }
}printf("SUCCESS\n");
```

★★

Code Snippet 8.3: The 1500 in the XBee_atModeOn argument is the time required to expire before placing the XBee module in AT command mode. A "1" is returned if the xb_atModeOn function completes successfully.

Just for grins, let's execute a bunch of XBee library functions and see what they do. I've captured the results of the execution of the functions in Code Snippet 8.4 as well.

Code Snippet 8.4

★★

```
xb_getCH();   // get the channel number
_atCmdRsp: Tx=ATCH
_atCmdRsp: Rx=B

xb_setDH(0); // set destination to 0 so no API's packets are sent out
_atCmdRsp: Tx=ATDH 00000000
_atCmdRsp: Rx=OK

xb_setDL(0); // set destination to 0 so no API's packets are sent out
_atCmdRsp: Tx=ATDL 00000000
_atCmdRsp: Rx=OK

xb_getMY();
_atCmdRsp: Tx=ATMY
_atCmdRsp: Rx=0

xb_getSH();
_atCmdRsp: Tx=ATSH
_atCmdRsp: Rx=13A200
```

```
xb_getSL();
_atCmdRsp: Tx=ATSL
_atCmdRsp: Rx=4008DD8D

xb_getRN();
_atCmdRsp: Tx=ATRN
_atCmdRsp: Rx=0

xb_setNI("DIO-COORD\r");   // setup a node id
_atCmdRsp: Tx=ATNI DIO-COORD
_atCmdRsp: Rx=OK

xb_getVR();
_atCmdRsp: Tx=ATVR
_atCmdRsp: Rx=10A1

xb_getHV();
_atCmdRsp: Tx=ATHV
_atCmdRsp: Rx=1706


// setup I/O for the RF Module Interface Board
xb_setD0(5); // out high
_atCmdRsp: Tx=ATD0 05
_atCmdRsp: Rx=OK

xb_setD1(5); // out high
_atCmdRsp: Tx=ATD1 05

_atCmdRsp: Rx=OK


xb_setD2(3); // S1 pushbutton
_atCmdRsp: Tx=ATD2 03
_atCmdRsp: Rx=OK

xb_setD3(2); // ADC3 for battery monitoring
_atCmdRsp: Tx=ATD3 02
_atCmdRsp: Rx=OK

xb_setD4(3); // S2 pushbutton
_atCmdRsp: Tx=ATD4 03
_atCmdRsp: Rx=OK

xb_atModeOff();
_atCmdRsp: Tx=ATCN
_atCmdRsp: Rx=OK
```

****

Code Snippet 8.4: The _atCmdRsp lines are showing what actually is being offered up on the serial port.

Note that in Code Snippet 8.4 we are matching up the PAN Coordinator's RF Module interface board configuration to the application's hardware configuration using a series of xb_setDX function calls. We also could have foregone using X-CTU to assign a Node Identifier as the xb_setNI function call in Code Snippet 8.4 does that for us.

The next code sequence shown in Code Snippet 8.5 is exclusive to XBee modules operating in a network. The xb_getND function fires off the data sequence I captured in Sniffer Capture 8.1.

Code Snippet 8.5
****

```
int   rval,samples,chi,dio,adc;
char data[1024]; // must be large enought to hold all discovered nodes
char *ptr;

        printf("Discovering Nodes... \n");
        xb_atModeOn(1500);
        waitfor((rval=xb_getND(data)));
        if(rval>0)
        {   printf(" Found nodes:\n");
         ptr = strtok(data,"\r"); // first call to strtok needs buffer
         while(ptr != NULL)
           {   printf("   MY: %s",ptr);
               printf("   SH: %s",strtok(NULL,"\r"));
               printf("   SL: %s",strtok(NULL,"\r"));
               printf("   DB: %s",strtok(NULL,"\r"));
               printf("   NI: %s",strtok(NULL,"\r"));
               printf("\n");
               ptr = strtok(NULL,"\r");   // see if there is another node
           }
        printf("End\n");
```

****
Code Snippet 8.5: The ND (Node Discover) function searches the network for XBee modules and if found returns their short address, their 64-bit IEEE address, the signal strength and their Node Identifier.

Naturally, the IEEE 802.15.4 stuff in Sniffer Capture 8.1 makes sense as the XBee modules are IEEE 802.15.4-compliant. The data in the MAC Payload is definitely proprietary as I could not locate any reference to the byte sequence anywhere in the XBee documentation.

Sniffer Capture 8.1
****

```
Frame 1 (Length = 24 bytes)
        Time Stamp: 15:12:19.000
        Frame Length: 24 bytes
        Capture Length: 24 bytes
        Link Quality Indication: 168
IEEE 802.15.4
        Frame Control: 0xc841
                .... .... .... .001  = Frame Type: Data (0x0001)
                .... .... .... 0...  = Security Enabled: Disabled
```

```
                .... .... ...0 ....   = Frame Pending: No more data
                .... .... ..0. ....   = Acknowledgment Request: Acknowledgment
not required
                .... .... .1.. ....   = Intra PAN: Within the PAN
                .... ..00 0... ....   = Reserved
                .... 10.. .... ....   = Destination Addressing Mode: Address
field contains a 16-bit short address (0x0002)
                ..00 .... .... ....   = Reserved
                11.. .... .... ....   = Source Addressing Mode: Address field
contains a 64-bit extended address (0x0003)
        Sequence Number: 252
        Destination PAN Identifier: 0xaaaa
        Destination Address: 0xffff
        Source Address: 0x0013a2004008dd8d
        Frame Check Sequence: Correct
MAC Payload: 05:01:00:01:b5:49:19

0000:   41 c8 fc aa aa ff ff 8d dd 08 40 00 a2 13 00 05   AH|**...].@."...
0010:   01 00 01 b5 49 19 .. ..                           ...5I...
```
****

Sniffer Capture 8.1: By letting the code cycle through the Node Discover function, I found that only the very first byte of the MAC Payload incremented every other transmission.

When a node is discovered, it associates with the PAN Coordinator in the standard IEEE 802.15.4 fashion and ships back the data package I captured in Sniffer Capture 8.2.

Sniffer Capture 8.2
****

```
Frame 27 (Length = 52 bytes)
        Time Stamp: 15:13:29.267
        Frame Length: 52 bytes
        Capture Length: 52 bytes
        Link Quality Indication: 176
IEEE 802.15.4
        Frame Control: 0xcc61
                .... .... .... .001   = Frame Type: Data (0x0001)
                .... .... .... 0...   = Security Enabled: Disabled
                .... .... ...0 ....   = Frame Pending: No more data
                .... .... ..1. ....   = Acknowledgment Request: Acknowledgment
required
                .... .... .1.. ....   = Intra PAN: Within the PAN
                .... ..00 0... ....   = Reserved
                .... 11.. .... ....   = Destination Addressing Mode: Address
field contains a 64-bit extended address (0x0003)
                ..00 .... .... ....   = Reserved
                11.. .... .... ....   = Source Addressing Mode: Address field
contains a 64-bit extended address (0x0003)
        Sequence Number: 50
        Destination PAN Identifier: 0xaaaa
        Destination Address: 0x0013a2004008dd8d
        Source Address: 0x0013a2004008dd58
```

```
        Frame Check Sequence: Correct
MAC Payload: 01:81:00:81:b5:c9:ff:fe:00:13:a2:00:40:08:dd:58:28:53:54:41:52:
43:48:49:4c:44:2d:31:00

0000:    61 cc 32 aa aa 8d dd 08 40 00 a2 13 00 58 dd     aL2**.].@.".X]
000f:    08 40 00 a2 13 00 01 81 00 81 b5 c9 ff fe 00     .@.".....5I.~.
001e:    13 a2 00 40 08 dd 58 28 53 54 41 52 43 48 49     .".@.]X(STARCHI
002d:    4c 44 2d 31 00 .. ..                             LD-1...
*************************************************************************
```

Sniffer Capture 8.2: The end device's Node Identifier is obvious in the hex dump portion of this capture. The MAC Payload, which contains the end-device information, begins at offset 0x0015.

Let's see if we can figure out what is going on here. Here's the Dynamic C STDIO printout:

```
*************************************************************************
 Discovering Nodes...
   Found nodes:
   MY: FFFE  SH: 13A200  SL: 4008DD58  DB: 28  NI: STARCHILD-1
 End
*************************************************************************
```

And, here's the hex dump of the data array gleaned from the response frame sent by the XBee End Device:

```
*************************************************************************
dbf2:  46 46 46 45 0D 31 33 41 32 30 30 0D 34 30 30 38    FFFE□13A200□4008
dc02:  44 44 35 38 0D 32 43 0D 53 54 41 52 43 48 49 4C    DD58□2C□STARCHIL
dc12:  44 2D 31 0D 00 00 00 00 00 00 00 00 00 00 00 00    D-1□
*************************************************************************
```

We can once again hang the proprietary tag on the first six bytes of the MAC Payload as they have no meaning to us at this point. All of the data fields in the data array are delimited by a carriage-return character (0x0D- "\r"). The strtok (string token) function in Code Snippet 8.5 is used to parse the data fields of the data array using the carriage return as the delimiter. The only piece of data that may not be obvious is the signal-strength value, which is converted to decibels before being output to the Dynamic C STDIO debugging window. A NULL (0x00) indicates the end of an end device's data structure. Multiple end-device data structures can be held in the data array, which is allocated as 1024 bytes.

Now, let's look at the code that produced the PAN Coordinator input samples. The sequence of events taking place in Code Snippet 8.6 work on the general-purpose I/O set-up we programmed into the PAN Coordinator XBee node in Code Snippet 8.4.

Code Snippet 8.6

```
*************************************************************************
        printf("Forcing input samples for the local XBee... \n");
        xb_getIS(data);       // force sample, get ADC
        samples = axtoi(strtok(data,"\r"));
        chi = axtoi(strtok(NULL,"\r"));
        dio = axtoi(strtok(NULL,"\r"));
        adc = axtoi(strtok(NULL,"\r"));
          printf("  samples(%04X)  channel Indicator(%04X)  active IOs(%04X)
```

```
                ADC3(%04X)\n\n",samples,chi,dio,adc);
                xb_atModeOff();
```
****
Code Snippet 8.6: This code really has nothing to do with IEEE 802.15.4 or ZigBee. Its purpose is to demonstrate that the XBee module can do more than just participate in an IEEE 802.15.4 PAN.

Here's what the data gathered from the PAN Coordinator's general-purpose I/O pins looks like in Rabbit microcontroller memory:
****
```
dbf2: 31 0D 31 30 31 37 0D 30 30 30 0D 31 45 37 0D 00    1□1017□000□1E7□
```
****

And, here's the resultant printout in the Dynamic C STDIO debugging window:
****
```
 Forcing input samples for the local XBee...
   samples(0001)  channel Indicator(1017)  active I/Os(0000)  ADC3(01E7)
```
****

You can readily see the relationship between the data fields in the hex dump, the argument fields of the printf function in Code Snippet 8.6 and the Dynamic C STDIO printout. This data was not presented to the RF portion of the XBee and therefore there's no Daintree Networks SNA capture to show. So, let's figure out how the data in the Dynamic C STDIO window came to be.

The first two data indicators are easily explained with a look at Figure 8.1.

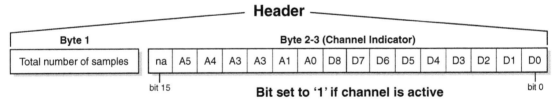

Figure 8.2: Nothing to it. Just match up the bits in the channel indicator to the bits in this figure.

If you simply match up 0x1017, which is the channel indicator value, to the bit layout in Figure 8.1 you'll find that in Code Snippet 8.4 we actually defined and set up every one of the active general-purpose I/O channels in our code.

Pushbutton switches on the XBee RF Module interface board are connected to DIO lines D2 and D4. Thus far, I have pressed no buttons as the active I/O's value is equal to 0 (zero).

To provide you with a better example of how the DIO fields work, I captured a session in which I depressed the S1 and S2 pushbuttons on the XBee RF Module interface board respectively. Here's what the Dynamic C STDIO window showed:
****
```
Discovering Nodes...
   Found nodes:
   MY: FFFE  SH: 13A200  SL: 4008DD58  DB: 32  NI: STARCHILD-1
```

```
End
Forcing input samples for the local XBee...
    samples(0001)  channel Indicator(1017)  active I/Os(0004)  ADC3(01E7)

Discovering Nodes...
    Found nodes:
    MY: FFFE  SH: 13A200  SL: 4008DD58  DB: 43  NI: STARCHILD-1
End
Forcing input samples for the local XBee...
    samples(0001)  channel Indicator(1017)  active I/Os(0010)  ADC3(01E6)
**************************************************************************
```

If you match up the 0x0004 with the DIO layout in Figure 8.2 and then correlate that back to the code in Code Snippet 8.4, you'll see that I was holding down the S1 pushbutton, which is tied to DIO2. Depressing S2 produced an active I/O value of 0x0010 that directly relates to DIO4 in Figure 8.2.

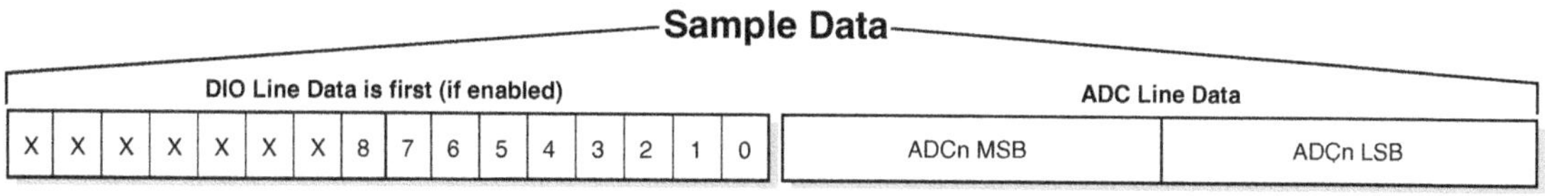

Figure 8.3: Again, nothing to it. The analog-to-digital converter value speaks for itself here.

It's pretty obvious that the firmware loaded on the XBee modules is an official IEEE 802.15.4 MAC and PHY implementation rolled in with some proprietary XBee functionality. The XBee modules are designed to work right out of the box and so are the Dynamic C-backed Rabbit RF interface modules. Although it is always good to have some background knowledge, no prior IEEE 802.15.4 or ZigBee experience is necessary to assemble a PAN using XBee modules and Rabbit microcontrollers.

Rabbit Semiconductor

Rabbit Semiconductor is a fabless semiconductor company specializing in high-performance 8-bit microprocessors and development tools for embedded control, communications, and Ethernet connectivity. I actually cut my adult Ethernet teeth with a Rabbit Ethernet development kit. A sister division of single-board computer and software manufacturer Z·World (I've done a few Z·World columns in my time), Rabbit Semiconductor introduced the Rabbit 2000® microprocessor in 1999 and the Rabbit 3000® in 2002 (I was all over these microprocessors). You can see all of my column links on the Rabbit Semiconductor site but all of the documents behind them are AWOL. Hmmmm…That's OK. As long as Rabbit Semiconductor and Z·World keep producing products, I'll keep on writing about them.

George Harrison played a variety of guitars during his stint as a Beatle, including a solid rosewood Fender Telecaster. You probably remember him chiming on a Rickenbacker 360 12-string. What you probably don't remember is George's 1962 Chet Atkins "Country Gentleman," which he favored on stage.

The next chapter will take us down South to Duluth, Georgia where Tim Cutler and the folks at Cirronet perform their IEEE 802.15.4/ZigBee magic.

Hey, hey! Don't turn that page yet. Since we're on our way to Georgia, do you know the name of the band that hails from Athens, Georgia and is considered the world's greatest party band? Hint: The frontman's name is Fred and one of the female singers has my wife's first name of Cindy.

Cirronet Adds Southern Flavor to IEEE 802.15.4 and ZigBee

One of the biggest pushes behind ZigBee is industrial use. There is nothing in the Zig-Bee specification that states that a ZigBee radio can't effectively radiate more than 1 mW. What I'm saying is, don't confuse low-power operation with the radiated power of an IEEE 802.15.4-compliant radio. A previous example of this line of thought is the XBee-Pro RF module, which can warp the electromagnetic surroundings with 100 mW of effective radiated power versus the 1 mW of power radiated by the standard XBee RF module. Tim Cutler and the folks at Cirronet are also part of the high-power industrial ZigBee congregation. Let's take a look at the Cirronet hardware we're about to examine.

We happen to have a Cirronet ZN241Z, which is factory programmed to act as a ZigBee Coordinator. As you can see in Photo 9.1, the ZN241Z is a ruggedized version of the Cirronet ZMN2400HP high-power IEEE 802.15.4-compliant radio.

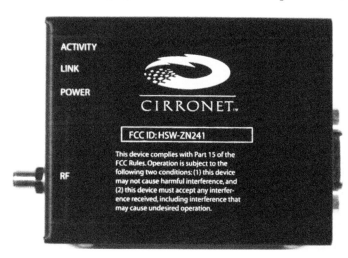

Photo 9.1: Cirronet ZN241Z. This baby is ready to rock and roll on the shop floor. A serial port connector and a mains power receptacle are mounted on the right quarterpanel.

I opened up the ZN241Z in Photo 9.2. What we have here is essentially a ZMN2400HP transceiver mounted on a motherboard with a hefty regulated power supply and a serial interface. The ZMN2400HP is designed to be soldered into place just like an IC. The IC comparison follows though as to how the ZMN2400HP is used as well.

There is no external access to the ZN241Z's ZMN2400HP radio other than the RF output and the serial port. That's fine as this module is intended to be the ZigBee Coordinator. Any important data transfers involving the ZN241Z can be accessed via the serial port. Note also the absence of any battery receptacles, which adheres to the thought of powering a ZigBee Coordinator from a mains supply.

We have another two-man PAN here. The device you see in Photo 9.3 is capable of being either a router or end device, which means it can operate as an FFD or an RFD.

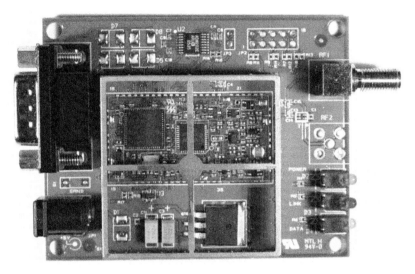

Photo 9.2: I removed the top cage so you could get a look at the Cirronet ZMN2400HP and the voltage regulator. The RS-232 converter IC is mounted top center of this shot. Power, Link and Data indicator LEDs are mounted at the lower right quarter.

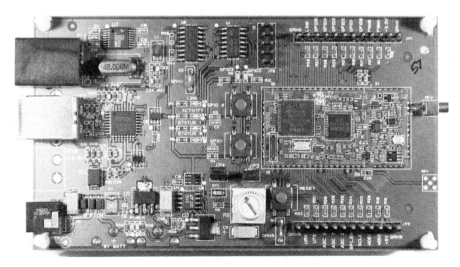

Photo 9.3: The idea behind this development board is to give the ZMN2400HP all of the necessary basic support it needs to operate, along with some sensing devices, which are tied to the ZMN2400HP's on-board Atmel ATmega128L.

The ZMN2400HP development board uses FTDI logic to emulate a serial port via USB. There is also a true RS-232 port available, which is accessed by a standard 6-pin RJ-11 jack. The USB and RS-232 ports are mutually exclusive in terms of usage. Power to the ZMN2400HP development board can be supplied by a mains wall wart or by way of a 9V battery, which snaps in on the opposite side of the board.

All of the ZMN2400HP's general-purpose I/O pins with the exception of RESET are pinned out along both sides of the ZMN2400HP transceiver module. For demonstration purposes, the ZMN2400HP development board is also fitted with a thermistor and potentiometer, which are tied to the ZMN2400HP's ADCX and ADCY lines, respectively. Jumpers are in place to allow the ZMN2400HP programmer to disconnect the on-board thermistor and potentiometer so that external devices can be wired into the ZMN2400HP's general-purpose I/O. A pair of active-low pushbutton switches is also available. The Atmel ATmega128L is capable of being debugged and programmed via its JTAG interface. To that end, the ZMN2400HP development board includes a standard ATmega128L JTAG interface that can be activated for special programming needs.

A bird's-eye view of the ZMN2400HP sits within the four corners of Photo 9.4. As you have already logically deduced, the ZMN2400HP's general-purpose I/O is actually Atmel ATmega128L general-purpose I/O. The Cirronet ZigBee stack code resides in the ATmega128L's program Flash and the RF duties are performed by a Texas Instruments/Chipcon CC2420.

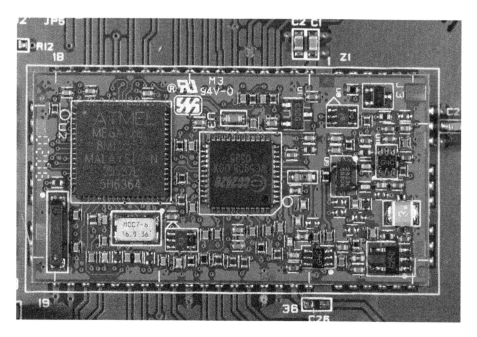

Photo 9.4: That discrete balun is a bit hard to pull from this shot. That's because there are quite a few more components between the CC2420's RF output pins and the actual antenna connector.

I initially fired up the ZN241Z ZigBee Coordinator module and looked for any signs of life with the Daintree Networks SNA. After a couple of tries, I switched channels. I finally found some life on channel 12. Again, to keep things under control, I connected the ZN241Z's serial port to my personal computer and kicked off the ZBDemo application. My intent was to set the operating channel to channel 11. I did just that in Screen Capture 9.1. I followed suit with the ZMN2400HP development board using the same process you see in Screen Capture 9.1 to select only channel 11.

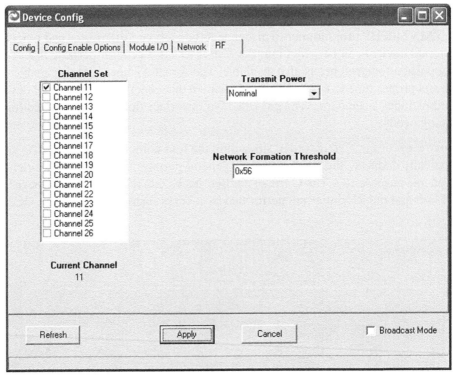

Screen Capture 9.1: This window is accessed via the Config button on the main ZBDemo

After making the channel changes, I refired the ZigBee Coordinator module and then applied some power to the ZMN2400HP development board. The ZigBee Coordinator performed an active scan that I saw as a Beacon Request via Daintree Networks SNA and established a PAN. I know a PAN was established because the ZMN2400HP development board associated with the ZigBee Coordinator module. I fiddled with the ZMN2400HP development board's potentiometer and punched at the pair of pushbutton switches. I couldn't make heads or tails of what was going on via the Daintree Networks SNA. So, I figured I'd change the configuration a bit. Maybe that would throw me a bone.

The ZMN2400HP development board comes from the factory programmed to act as a ZigBee Router. A trio of programs is included on the ZMN24HPDK-Basic CD-ROM that allows the Cirronet transceiver modules to be programmed as a ZigBee Coordinator, a ZigBee Router or a

ZigBee End Device. A program to load the code into the ZMN2400HP's ATmega128L is also part of the code package. I opted to change the ZMN2400HP development board's role from a ZigBee Router to a ZigBee End Device. After a few tries, I found that the bootloader program would not work directly from the development kit CD-ROM. I transferred the bootloader program and the three ZigBee component files over to my personal computer's hard drive and successfully converted the ZMN2400HP development board to a ZigBee End Device. I again powered everything down and brought up the ZigBee Coordinator module first. The newly crowned ZigBee End Device associated as it should and began to transmit ZigBee data messages.

The key to interpreting the ZigBee messages flowing between the ZN241Z and ZMN2400HP lies in the understanding of the Cirronet Standard Module (CSM) Profile API. In the case of the Cirronet RF modules, the CSM profile is a Cirronet-specific application profile that defines the messaging scheme used by Cirronet ZigBee radios. A combination of the messaging elements contained with the CSM profile make up the ZigBee module's (the Cirronet radio module's) resident application. This puts the CSM profile in the application area of the ZigBee stack. In fact, you'll see that the Daintree Networks SNA decodes the CSM profile messages in the ZigBee APS and ZigBee AF areas.

Interestingly, the CSM profile used by the Cirronet ZigBee modules uses a serial packet message/command protocol to communicate through the module's UART or via RF messages. The protocol commands can be used to set configuration parameters, issue commands or transfer data. To keep it simple (the way I like it), all of the packets follow a common layout. The actual message contained within the packet is defined by a field within the packet. The generic packet format is depicted in Figure 9.1.

1 byte	1 byte	1 byte	1 byte	Varies
SOP (0xFD)	Length (in bytes)	TransID	MSG Type	Arguments

Figure 9.1: Generic packet format. The MSG Type byte is the key to what lies in the Arguments area.

The SOP (Start Of Packet) byte is a constant 0xFD. You'll find that 0xFD is an easy marker to find in hex dumps of these serial packets. The SOP byte is not counted in the packet-length byte that follows it.

When an ACK is not received in time by the sender of a packet, the sender will try up to seven times to resend it. Just because the ACK was not sensed in time does not always mean that the data was not delivered. So, to help keep these ACK disagreements at a minimum, the TransID field is included in the Cirronet serial packet format. It is recommended that the application auto-increment the TransID value to allow for the applications on both ends to figure out what data is new and what data is old in the data-acknowledgment cycles.

The fun begins with the MSG Type byte, which determines what type of operation is being performed or what type of data is being manipulated. Here's a look at all of the MSG Types used by the CSM profile:

- 0x01: Set Field

- 0x05: Get Field

- 0x0A: Send String

- 0x0C: Send SPI

- 0x10: Get IEEE Address

- 0x11: Get NWK Address

- 0x64: Discovery Request

- 0x65: Discovery End

- 0x81: Set Reply

- 0x85: Get Reply

- 0x8A: Send String Reply

- 0x8C: Send SPI Reply

- 0x8E: Receive String

- 0x90: Get IEEE Address Reply

- 0x91: Get NWK Address Reply

- 0x95: Receive Field Event

- 0xD0: Link Announce

- 0xE4: Discovery Reply

- 0xF0: Device Registration

- 0xFF: Error

The CSM profile packet ends with the Arguments field. Figure 9.2 shows how the subfields are defined within the Arguments field.

8 bytes	2 bytes	1 byte	1 byte	2 bytes	1 byte	Varies
MAC/NWK Address**	ProfileID	Endpoint	Cluster	Offset	Length	Data

Figure 9.2: Filling out these fields is easier than it looks. The ProfileID is a constant, the Endpoint has only one value and the Cluster is one of the aforementioned CSM profile clusters I listed for you.

The MAC/NWK address field is used for the standard IEEE 64-bit address of the short 16-bit address. Regardless of which address length is used, the MAC/NWK length of 64 bits is always honored. That keeps the packet format intact. Cirronet's IEEE-assigned address prefix is 00:30:66, which always puts a 0x00 in the most significant byte of the Cirronet 64-bit address. To distinguish a 16-bit address, the most significant byte of the MAC/NWK field must be set to 0x80 with the two least significant bytes of the MAC/NWK field containing the 16-bit address value. The remaining five bytes are set to 0x00.

Right now, the Cirronet ProfileID value is constant at 0xC000. The constant idea can also be applied to the Endpoint field, as the CSM profile only supports an endpoint of 0x01.

The CSM profile groups its operational parameters within a set of clusters, which are stored in NVRAM. The data elements and devices such as ADC X and ADC Y are addressed as offsets into the respective cluster that is associated with the data element or device. A cluster is a collection of memory locations containing configuration parameters. The functionality of a ZigBee device is defined by the contents of its clusters. Here's a list of the CSM profile clusters:

- Module I/O Cluster (ID 0x01)
- Configuration Cluster (ID 0x02)
- Reset Cluster (ID 0x03)
- Network Cluster (ID 0x07)
- RF Cluster (ID 0x08)
- Security Cluster (ID 0x09)

All of the cluster offsets begin at 0x0000. ADC X is a good example as its 2-byte value resides in the Module I/O Cluster at offset 0x0000. To gain access to the Cirronet transceiver's SPI portal, you would access offset 0x0010 in the Module I/O Cluster. The UART is at offset 0x0011. Get the idea?

The Configuration Cluster is the place for network set-up. You define the role of the Cirronet RF module (Coordinator, Router, End Device) by writing a 0x00, 0x01 or Cirronet 0x02 respectively to offset 0x0002 of the Configuration Cluster. Change the Cirronet ZMN2400HP's baud rate by writing a value to the Serial Mode parameter located at offset 0x0003 of the Configuration Cluster. The ZMN2400HP also incorporates what it calls a Friendly Name (as compared to the Node Identifier used by the XBee modules), which is a 16-byte ASCII value that can be stuffed into offset 0x0007 of the Configuration Cluster.

The Reset Cluster has only two offsets. Offset 0x0000 needs a value of 0x5A to enable the reset of the ZMN2400HP's ATmega128L microcontroller. Writing a 0x5A to offset 0x0001 of the Reset Cluster resets all of the cluster parameters to factory default values and then resets the ATmega128L microcontroller.

Want to see the factory-programmed MAC address? All you have to do is read the 8 bytes that make up the IEEE MAC address at offset 0x0000 of the Network Cluster. The PAN ID can be set by writing 2 bytes of PAN ID information to offset 0x0013.

Recall that I used the ZBDemo RF configuration window to set up the network for channel 11. The same function can be performed by entering a value in the Channel List parameter at offset 0x0000 of the RF Cluster. The current channel being used by the ZMN2400HP can be obtained from the value of offsef 0x0004.

I've not mentioned security and I won't start now. Depending on the documents you read about ZigBee security, you'll find that some of them declare it null and void while others offer up their views of how to apply it. I'll leave security and the application thereof up to you, as we won't be discussing it in the pages of this book.

Values are written to a cluster using the Set Field command. Conversely, Get Field commands retrieve data from cluster elements. As you have already deduced, the ZMN2400HP is controlled by Set Field commands. For instance, the Get Field command is used to harvest the values of the ZMN2400HP's general-purpose I/O lines from the Module I/O Cluster.

The Length byte of the packet layout in Figure 9.2 represents the number of bytes in the Data area that follows it. In a reply packet, the LQI value is included as an extra parameter.

The ZBDemo program is simply an interface that sends and receives the formatted serial packets to the ZMN2400HP it is connected to via a physical personal computer serial port. Tapping into the ZBDemo's serial link would be a great, albeit manual, way of examining the CSM profile API. However, it's bunches of more fun to dig through hex dumps and captures. So, I've employed Daintree Networks SNA to capture the data transfers between the Cirronet ZigBee Coordinator and ZigBee End Device. Let's bust up the Daintree Networks SNA ZigBee capture I've pulled out of the air in Sniffer Capture 9.1.

Sniffer Capture 9.1

```
*************************************************************************
Frame 21 (Length = 48 bytes)
        Time Stamp: 18:14:45.577
        Frame Length: 48 bytes
        Capture Length: 48 bytes
        Link Quality Indication: 216
IEEE 802.15.4
        Frame Control: 0x8861
                .... .... .... .001  = Frame Type: Data (0x0001)
                .... .... .... 0...  = Security Enabled: Disabled
                .... .... ...0 ....  = Frame Pending: No more data
                .... .... ..1. ....  = Acknowledgment Request: Acknowledgment
required
                .... .... .1.. ....  = Intra PAN: Within the PAN
                .... ..00 0... ....  = Reserved
                .... 10.. .... ....  = Destination Addressing Mode: Address
field contains a 16-bit short address (0x0002)
                ..00 .... .... ....  = Reserved
                10.. .... .... ....  = Source Addressing Mode: Address field
contains a 16-bit short address (0x0002)
        Sequence Number: 31
        Destination PAN Identifier: 0x307d
        Destination Address: 0x0000
```

```
        Source Address: 0x796f
        Frame Check Sequence: Correct
ZigBee NWK
        Frame Control: 0x0044
                .... .... .... ..00  = Frame Type: NWK Data (0x00)
                .... .... ..00 01..  = Protocol Version (0x01)
                .... .... 01.. ....  = Discover Route: Enable route discovery
(0x01)
                .... ...0 .... ....  = Reserved
                .... ..0. .... ....  = Security: Disabled
                0000 00.. .... ....  = Reserved
        Destination Address: 0x0000
        Source Address: 0x796f
        Radius = 7
        Sequence Number = 3
ZigBee APS
        Frame Control: 0x00
                .... ..00  = Frame Type: APS Data (0x00)
                .... 00..  = Delivery Mode: Normal Unicast Delivery (0x00)
                ...0 ....  = Indirect Address Mode: Ignored
                ..0. ....  = Security: False
                .0.. ....  = Ack Request: Acknowledgment not required
                0... ....  = Reserved
        Destination Endpoint: 0x01
        Cluster Identifier:  (0x0001)
        Profile Identifier:  (0xc000)
        Source Endpoint: 0x01
ZigBee AF
        AF Header: 0x21
                .... 0001  = Transaction Count: (0x01)
                0010 ....  = Frame Type: MSG (0x02)
        Transaction 1
                Transaction Sequence Number = 0x01
                ZigBee MSG
                        Transaction Length: 20
                        Transaction Data: fd:12:01:95:6f:79:00:00:00:00:00:80:0
0:c0:01:01:0b:00:01:00

0000:   61 88 1f 7d 30 00 00 6f 79 44 00 00 00 6f 79 07     a..}0..oyD...oy.
0010:   03 00 01 01 00 c0 01 21 01 14 fd 12 01 95 6f 79     .....@.!..}...oy
0020:   00 00 00 00 00 80 00 c0 01 01 0b 00 01 00 .. ..     .......@........
```
****

Sniffer Capture 9.1: Thanks to the ZMN2400HP's built-in ZigBee stack, we've reached out a couple of ZigBee stack layers above the MAC layer in this capture.

I've made this statement many times in the pages of this book. If you see something new in Sniffer Capture 9.2, sell this book or give it to your sister. I can tell you something about Sniffer Capture 9.2 you may not know. The Destination PAN Identifier is the default value, which is stored in the Network Cluster at offset 0x0013.

Sniffer Capture 9.2

```
**************************************************************************
Frame 21 (Length = 48 bytes)
        Time Stamp: 18:14:45.577
        Frame Length: 48 bytes
        Capture Length: 48 bytes
        Link Quality Indication: 216
IEEE 802.15.4
        Frame Control: 0x8861
                .... .... .... .001  = Frame Type: Data (0x0001)
                .... .... .... 0...  = Security Enabled: Disabled
                .... .... ...0 ....  = Frame Pending: No more data
                .... .... ..1. ....  = Acknowledgment Request: Acknowledgment
required
                .... .... .1.. ....  = Intra PAN: Within the PAN
                .... ..00 0... ....  = Reserved
                .... 10.. .... ....  = Destination Addressing Mode: Address
field contains a 16-bit short address (0x0002)
                ..00 .... .... ....  = Reserved
                10.. .... .... ....  = Source Addressing Mode: Address field
contains a 16-bit short address (0x0002)
        Sequence Number: 31
        Destination PAN Identifier: 0x307d
        Destination Address: 0x0000
        Source Address: 0x796f
        Frame Check Sequence: Correct
**************************************************************************
```

Sniffer Capture 9.2: Nothing new here. This is business as usual in the ZigBee/IEEE 802.15.4 world.

You've not seen many if any NWK sniffs. However, even though it is a necessary ZigBee component, it is uneventful in this scenario. The Radius value in Sniffer Capture 9.3 has to do with how many hops the message may be able to take in this network. Since we only have two nodes, you can talk about the Radius value amongst yourselves.

Sniffer Capture 9.3

```
**************************************************************************
ZigBee NWK
        Frame Control: 0x0044
                .... .... .... ..00  = Frame Type: NWK Data (0x00)
                .... .... ..00 01..  = Protocol Version (0x01)
                .... .... 01.. ....  = Discover Route: Enable route discovery
(0x01)
                .... ...0 .... ....  = Reserved
                .... ..0. .... ....  = Security: Disabled
                0000 00.. .... ....  = Reserved
        Destination Address: 0x0000
        Source Address: 0x796f
        Radius = 7
        Sequence Number = 3
**************************************************************************
```

Sniffer Capture 9.3: Repeat after me, "There are no hops in this two-man network..

Sniffer Capture 9.4 is the APS portion of the original Sniffer Capture 9.1. Keep in mind that the ZigBee Coordinator and ZigBee End Device applications are communicating with each other using their respective endpoints numbered 0x01.

The Cluster Identifier and Profile Identifier should not be strange concepts.

Sniffer Capture 9.4
```
**********************************************************************
ZigBee APS
      Frame Control: 0x00
                  .... ..00  = Frame Type: APS Data (0x00)
                  .... 00..  = Delivery Mode: Normal Unicast Delivery (0x00)
                  ...0 ....  = Indirect Address Mode: Ignored
                  ..0. ....  = Security: False
                  .0.. ....  = Ack Request: Acknowledgment not required
                  0... ....  = Reserved
      Destination Endpoint: 0x01
      Cluster Identifier:  (0x0001)
      Profile Identifier:  (0xc000)
      Source Endpoint: 0x01
**********************************************************************
```
Sniffer Capture 9.4: No brainer. The CSM profile is identified by 0xC000 and it only supports endpoint 0x01.

This is where the serial packet payload resides. However, we have to dig just a bit farther to get at the data. The first byte of Sniffer Capture 9.5 should be calling your name.

Sniffer Capture 9.5
```
**********************************************************************
ZigBee AF
      AF Header: 0x21
                  .... 0001  = Transaction Count: (0x01)
                  0010 ....  = Frame Type: MSG (0x02)
      Transaction 1
            Transaction Sequence Number = 0x01
            ZigBee MSG
                  Transaction Length: 20
                  Transaction Data: fd:12:01:95:6f:79:00:00:00:00:00:80:0
0:c0:01:01:0b:00:01:00
**********************************************************************
```
Sniffer Capture 9.5: A ZigBee message doesn't have to follow any standard guidelines. All that is necessary is that both ends of the conversation understand each other.

Let's break down the bytes in the Transaction Data. The 0xFD is a given as it is the SOP byte. There are 18 bytes (0x12) that follow in the packet. The application has assigned a TransID of 0x01.

Now, here's where it gets interesting. The MSG Type is 0x95, which is defined in the CSM profile documentation as a Receive Field Event. A Receive Field Event signals a message that is not generated by a field device in response to a Coordinator-initiated message. In this case I pressed a pushbutton on the ZMN2400HP development board that prompted the transmission of the general-purpose I/O change. I've posted the Receive Field Event byte layout in Figure 9.3.

Arguments (Receive Field Event)						
MAC Address (8 bytes)	ProfileID (2 bytes)	Endpoint (1 byte)	Cluster (1 byte)	Offset (2 bytes)	Length (1 byte)	[Data] (LSB First)

Figure 9.3: Receive Field Event byte layout. You're already stuffing those Transaction Data bytes into these fields, aren't you?

We've been told in the sniff that 16-bit addresses are in use. That means that the most significant byte of the MAC Address field will be 0x80 and the 16-bit address will reside at the least-significant pair of bytes. Note that the MAC Address bytes are arranged in least-significant-byte-first order. In other words, read backwards 8 bytes from the 0x80 byte to get the MAC Address value.

The ProfileID should be a no-brainer as it can only assume the value of 0xC000 as dictated by the CSM profile. The same duh-huh (that's Southern for uh-huh) goes for the Endpoint value, which can only be 0x01.

Cluster 0x01 does indeed have an offset of 0x000B, which happens to be designated as the GP I/O 1 parameter. GP I/O 1 represents the ZMN2400HP pin that is connected to the pushbutton I put my big fat finger on. The length of the data payload is 1 byte and that data byte is 0x00, which says the pushbutton is depressed presenting a logical low to the ZMN2400HP's GP I/O 1.

Let's generate some ZigBee fun of our own. The Cirronet ZMN2400HP development package also comes with a personal computer-based program called WinCom. We'll use WinCom to assemble and transmit a Discovery Request message and a Get IEEE Address message. This is merely an exercise to demonstrate how the CSM profile API functions. You already know that during the association process all of the data we're gathering has already been passed between the PAN Coordinator and the end device. The messages we're about to generate are typically used to discover unknown Cirronet nodes that are capable of communication in the ZigBee Coordinator's operating space. This could come in handy if one needed to gather a list of active Cirronet nodes for management purposes.

The Discovery Request message looks for any device within radio range with a profile that matches the Coordinator's profile. If such a device is present and hears the request, it will respond with a Discovery Reply message containing its network address. The ZigBee Coordinator can then use the newly found network address to obtain the remote device's IEEE address with the Get IEEE Address message.

The layout of the Discovery Request message lies in Figure 9.4. The only entry you really have to think about is the Timeout value, which is entered in seconds.

Arguments (Discovery Request)		
ProfileID (2 bytes)	Endpoint (1 byte)	Timout (1 byte)

Figure 9.4: Hmmmm...Let's see, hexadecimal 00C0 01 03 sounds good to me.

Screen Capture 9.2 is a view of the WinCom window I used to generate the Discovery Request message.

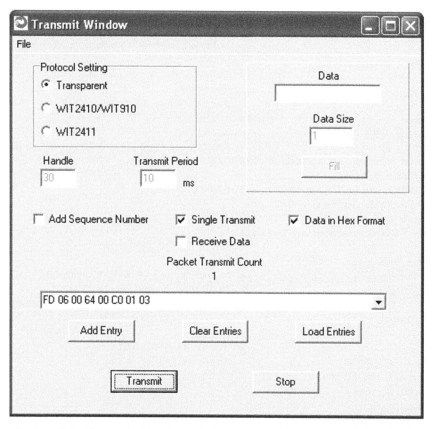

Screen Capture 9.2: The packet I entered follows the CSM profile's serial packet layout and includes the argument list for the Discovery Request message.

The 0xFD tells us that this is a CSM profile serial packet. Six bytes will follow the length byte with the first byte, 0x00, being the transaction ID (TransID). The Discovery Request arguments require the CSM profile ProfileID of 0xC000 to be followed by the CSM profile-supported Endpoint of 0x01. Just for grins, we'll set the Timeout value at 0x03 seconds. That simply means that after 3 seconds, the ZigBee Coordinator that generates the Discovery Request message will cease to accept Discovery Reply messages. You can bet your booty that I'm running the Daintree Networks SNA application and the content of Sniffer Capture 9.6 are what I got after clicking the Transmit button.

Sniffer Capture 9.6

```
**************************************************************************
Frame 1 (Length = 52 bytes)
        Time Stamp: 10:20:21.000
        Frame Length: 52 bytes
        Capture Length: 52 bytes
        Link Quality Indication: 220
IEEE 802.15.4
        Frame Control: 0x8841
                .... .... .... .001 = Frame Type: Data (0x0001)
                .... .... .... 0... = Security Enabled: Disabled
                .... .... ...0 .... = Frame Pending: No more data
                .... .... ..0. .... = Acknowledgment Request: Acknowledgment
not required
                .... .... .1.. .... = Intra PAN: Within the PAN
                .... ..00 0... .... = Reserved
                .... 10.. .... .... = Destination Addressing Mode: Address
field contains a 16-bit short address (0x0002)
                ..00 .... .... .... = Reserved
                10.. .... .... .... = Source Addressing Mode: Address field
contains a 16-bit short address (0x0002)
        Sequence Number: 78
        Destination PAN Identifier: 0x307d
        Destination Address: 0xffff
        Source Address: 0x0000
        Frame Check Sequence: Correct
ZigBee NWK
        Frame Control: 0x0004
                .... .... .... ..00 = Frame Type: NWK Data (0x00)
                .... .... ..00 01.. = Protocol Version (0x01)
                .... .... 00.. .... = Discover Route: Suppress route
discovery (0x00)
                .... ...0 .... .... = Reserved
                .... ..0. .... .... = Security: Disabled
                0000 00.. .... .... = Reserved
        Destination Address: 0xffff
        Source Address: 0x0000
        Radius = 7
        Sequence Number = 17
ZigBee APS
        Frame Control: 0x08
                .... ..00 = Frame Type: APS Data (0x00)
                .... 10.. = Delivery Mode: Broadcast (0x02)
                ...0 .... = Indirect Address Mode: Ignored
                ..0. .... = Security: False
                .0.. .... = Ack Request: Acknowledgment not required
                0... .... = Reserved
        Destination Endpoint: 0x00
        Cluster Identifier: MatchDescReq (0x0006)
        Profile Identifier: ZDP (0x0000)
        Source Endpoint: 0x00
```

```
ZigBee AF
      AF Header: 0x21
            .... 0001  = Transaction Count: (0x01)
            0010 ....  = Frame Type: MSG (0x02)
      Transaction 1
            Transaction Sequence Number = 0x0c
            ZigBee MSG
                  Transaction Length: 24
                  Transaction Data: ff:ff:00:c0:09:01:02:03:04:07:08:09:0
5:06:09:01:02:03:04:07:08:09:05:06
            ZigBee ZDO
                  NWK Address Of Interest: 0xffff
                  Profile Id:  (0xc000)
                  Number of Input Clusters: 9
                  Input Cluster List
                        Cluster Identifier 1:   (0x01)
                        Cluster Identifier 2:   (0x02)
                        Cluster Identifier 3:   (0x03)
                        Cluster Identifier 4:   (0x04)
                        Cluster Identifier 5:   (0x07)
                        Cluster Identifier 6:   (0x08)
                        Cluster Identifier 7:   (0x09)
                        Cluster Identifier 8:   (0x05)
                        Cluster Identifier 9:   (0x06)
                  Number of Output Clusters: 9
                  Output Cluster List
                        Cluster Identifier 1:   (0x01)
                        Cluster Identifier 2:   (0x02)
                        Cluster Identifier 3:   (0x03)
                        Cluster Identifier 4:   (0x04)
                        Cluster Identifier 5:   (0x07)
                        Cluster Identifier 6:   (0x08)
                        Cluster Identifier 7:   (0x09)
                        Cluster Identifier 8:   (0x05)
                        Cluster Identifier 9:   (0x06)

0000:   41 88 4e 7d 30 ff ff 00 00 04 00 ff ff 00 00 07
A.N}0...........
0010:   11 08 00 06 00 00 00 21 0c 18 ff ff 00 c0 09 01
.......!.....@..
0020:   02 03 04 07 08 09 05 06 09 01 02 03 04 07 08 09  ...............
0030:   05 06 .. ..                                        ....
****************************************************************************
```

Sniffer Capture 9.6: This is a ZigBee MatchDescReq (Match Description Request) message. Note that it is not a CSM profile API call as the cluster identifier is not one of the CSM profile clusters. That's OK. That's why we have Daintree Networks SNA.

The physicals of the ZigBee PAN must be satisfied no matter what we do. So, note the important clues given to you in Sniffer Capture 9.7.

Sniffer Capture 9.7

```
*********************************************************************************
Frame 1 (Length = 52 bytes)
        Time Stamp: 10:20:21.000
        Frame Length: 52 bytes
        Capture Length: 52 bytes
        Link Quality Indication: 220
IEEE 802.15.4
        Frame Control: 0x8841
                .... .... .... .001  = Frame Type: Data (0x0001)
                .... .... .... 0...  = Security Enabled: Disabled
                .... .... ...0 ....  = Frame Pending: No more data
                .... .... ..0. ....  = Acknowledgment Request: Acknowledgment
not required
                .... .... .1.. ....  = Intra PAN: Within the PAN
                .... ..00 0... ....  = Reserved
                .... 10.. .... ....  = Destination Addressing Mode: Address
field contains a 16-bit short address (0x0002)
                ..00 .... .... ....  = Reserved
                10.. .... .... ....  = Source Addressing Mode: Address field
contains a 16-bit short address (0x0002)
        Sequence Number: 78
        Destination PAN Identifier: 0x307d
        Destination Address: 0xffff
        Source Address: 0x0000
        Frame Check Sequence: Correct
*********************************************************************************
```

Sniffer Capture 9.7: This is a broadcast message emanating for the 0x307D ZigBee PAN that was generated by the ZigBee PAN Coordinator at address 0x0000.

Once the physical ZigBee components are inline, the next step up the stack must be satiated. Sniffer Capture 9.8 represents the logical addressing needs of the NWK layer of the ZigBee stack. Note the logical NWK sequence number in Sniffer Capture 9.8 is not at all related to the physical IEEE 802.15.4 sequence number in Sniffer Capture 9.7.

Sniffer Capture 9.8

```
*********************************************************************************
ZigBee NWK
        Frame Control: 0x0004
                .... .... .... ..00  = Frame Type: NWK Data (0x00)
                .... .... ..00 01..  = Protocol Version (0x01)
                .... .... 00.. ....  = Discover Route: Suppress route
discovery (0x00)
                .... ...0 .... ....  = Reserved
                .... ..0. .... ....  = Security: Disabled
                0000 00.. .... ....  = Reserved
        Destination Address: 0xffff
        Source Address: 0x0000
        Radius = 7
        Sequence Number = 17
*********************************************************************************
```

Sniffer Capture 9.8: You didn't think I would let you go without telling you what the Radius parameter really is, did you? Officially, Radius is defined as the distance, in hops, that a frame will be allowed to travel through the network.

Moving on up the George Jefferson way brings us to the next level of the ZigBee stack, the APS (Application Support Sublayer). Let's use what we know to figure out what the ZigBee APS sniff in Sniffer Capture 9.9 is telling us.

Sniffer Capture 9.9

```
*************************************************************************
ZigBee APS
        Frame Control: 0x08
                .... ..00  = Frame Type: APS Data (0x00)
                .... 10..  = Delivery Mode: Broadcast (0x02)
                ...0 ....  = Indirect Address Mode: Ignored
                ..0. ....  = Security: False
                .0.. ....  = Ack Request: Acknowledgment not required
                0... ....  = Reserved
        Destination Endpoint: 0x00
        Cluster Identifier: MatchDescReq (0x0006)
        Profile Identifier: ZDP (0x0000)
        Source Endpoint: 0x00
*************************************************************************
```

Sniffer Capture 9.9: Knowing simple and basic things leads to understanding large and complex things.

Let's begin by identifying the Profile Identifer. ZDP is short for ZigBee Device Profile. The active elements of the ZDP interact directly with the ZigBee stack's ZDO (ZigBee Device Objects). The ZDO is accessed exclusively via endpoint 0x00. The Cluster Identifier in Sniffer Capture 9.9, 0x0006, indeed represents the Match_Desc_req primitive called out in the ZigBee specification. The Match_Desc_req is generated from a local device wishing to find remote devices that have matching criteria representative of that listed in Sniffer Capture 9.10.

Sniffer Capture 9.10

```
*************************************************************************
ZigBee AF
        AF Header: 0x21
                .... 0001  = Transaction Count: (0x01)
                0010 ....  = Frame Type: MSG (0x02)
        Transaction 1
                Transaction Sequence Number = 0x0c
                ZigBee MSG
                        Transaction Length: 24
                        Transaction Data:
ff:ff:00:c0:09:01:02:03:04:07:08:09:05:06:09:01:02:03:04:07:08:09:05:06
                ZigBee ZDO
                        NWK Address Of Interest: 0xffff
                        Profile Id:  (0xc000)
                        Number of Input Clusters: 9
                        Input Cluster List
```

```
                                 Cluster Identifier 1:   (0x01)
                                 Cluster Identifier 2:   (0x02)
                                 Cluster Identifier 3:   (0x03)
                                 Cluster Identifier 4:   (0x04)
                                 Cluster Identifier 5:   (0x07)
                                 Cluster Identifier 6:   (0x08)
                                 Cluster Identifier 7:   (0x09)
                                 Cluster Identifier 8:   (0x05)
                                 Cluster Identifier 9:   (0x06)
                    Number of Output Clusters: 9
                    Output Cluster List
                                 Cluster Identifier 1:   (0x01)
                                 Cluster Identifier 2:   (0x02)
                                 Cluster Identifier 3:   (0x03)
                                 Cluster Identifier 4:   (0x04)
                                 Cluster Identifier 5:   (0x07)
                                 Cluster Identifier 6:   (0x08)
                                 Cluster Identifier 7:   (0x09)
                                 Cluster Identifier 8:   (0x05)
                                 Cluster Identifier 9:   (0x06)
```

****

Sniffer Capture 9.10: This is basically a list that the receiving device needs to match. Note the inclusion of the Cirronet CSM profile clusters plus a couple that probably come from within the ZDP.

The contents of Sniffer Capture 9.10 are reaching out to any remote device that can match up with its ProfileID and cluster list. That's going to be a given in our situation as the remote device is indeed a bonafide Cirronet ZigBee module, which answers accordingly in Sniffer Capture 9.11.

Sniffer Capture 9.11
****

```
Frame 2 (Length = 33 bytes)
        Time Stamp: 10:20:21.007
        Frame Length: 33 bytes
        Capture Length: 33 bytes
        Link Quality Indication: 176
IEEE 802.15.4
        Frame Control: 0x8861
                    .... .... .... .001   = Frame Type: Data (0x0001)
                    .... .... .... 0...    = Security Enabled: Disabled
                    .... .... ...0 ....    = Frame Pending: No more data
                    .... .... ..1. ....    = Acknowledgment Request: Acknowledgment
required
                    .... .... .1.. ....    = Intra PAN: Within the PAN
                    .... ..00 0... ....    = Reserved
                    .... 10.. .... ....    = Destination Addressing Mode: Address
field contains a 16-bit short address (0x0002)
                    ..00 .... .... ....    = Reserved
                    10.. .... .... ....    = Source Addressing Mode: Address field
contains a 16-bit short address (0x0002)
        Sequence Number: 38
        Destination PAN Identifier: 0x307d
```

```
        Destination Address: 0x0000
        Source Address: 0x796f
        Frame Check Sequence: Correct
ZigBee NWK
        Frame Control: 0x0004
                .... .... .... ..00  = Frame Type: NWK Data (0x00)
                .... .... ..00 01..  = Protocol Version (0x01)
                .... .... 00.. ....  = Discover Route: Suppress route
discovery (0x00)
                .... ...0 .... ....  = Reserved
                .... ..0. .... ....  = Security: Disabled
                0000 00.. .... ....  = Reserved
        Destination Address: 0x0000
        Source Address: 0x796f
        Radius = 7
        Sequence Number = 10
ZigBee APS
        Frame Control: 0x40
                .... ..00  = Frame Type: APS Data (0x00)
                .... 00..  = Delivery Mode: Normal Unicast Delivery (0x00)
                ...0 ....  = Indirect Address Mode: Ignored
                ..0. ....  = Security: False
                .1.. ....  = Ack Request: Acknowledgment required
                0... ....  = Reserved
        Destination Endpoint: 0x00
        Cluster Identifier: MatchDescRsp (0x0086)
        Profile Identifier: ZDP (0x0000)
        Source Endpoint: 0x00
ZigBee AF
        AF Header: 0x21
                .... 0001  = Transaction Count: (0x01)
                0010 ....  = Frame Type: MSG (0x02)
        Transaction 1
                Transaction Sequence Number = 0x0c
                ZigBee MSG
                        Transaction Length: 5
                        Transaction Data: 00:6f:79:01:01
                ZigBee ZDO
                        Status = Success: (0x00)
                        NWK Address Of Interest: 0x006f
                        Match Length: 1
                        Match List
                                Endpoint 1: 0x01

0000:   61 88 26 7d 30 00 00 6f 79 04 00 00 00 6f 79 07    a.&}0..oy....oy.
0010:   0a 40 00 86 00 00 00 21 0c 05 00 6f 79 01 01 ..    .@.....!...oy...
0020:                                                       ..
.
```

****

Sniffer Capture 9.11: Everything up to the ZigBee AF sniff data should be old hat. Note that the Transaction Sequence Numbers match the ZigBee AF sniff request in Sniffer Capture 9.10 to the ZigBee AF sniff response in Sniffer Capture 9.11.

Let's cut to the chase. The ZigBee APS sniff data tells us that this frame is a Match_Desc_ resp. A status of 0x00 is returned indicating a successful request. The receiving Cirronet ZigBee node found itself in a matching situation and returned its 16-bit network address and matching endpoint (0x01) to the ZigBee Coordinator.

Sniffer Capture 9.12

```
**********************************************************************
ZigBee APS
        Frame Control: 0x40
                .... ..00  = Frame Type: APS Data (0x00)
                .... 00..  = Delivery Mode: Normal Unicast Delivery (0x00)
                ...0 ....  = Indirect Address Mode: Ignored
                ..0. ....  = Security: False
                .1.. ....  = Ack Request: Acknowledgment required
                0... ....  = Reserved
        Destination Endpoint: 0x00
        Cluster Identifier: MatchDescRsp (0x0086)
        Profile Identifier: ZDP (0x0000)
        Source Endpoint: 0x00
ZigBee AF
        AF Header: 0x21
                .... 0001  = Transaction Count: (0x01)
                0010 ....  = Frame Type: MSG (0x02)
        Transaction 1
                Transaction Sequence Number = 0x0c
                ZigBee MSG
                        Transaction Length: 5
                        Transaction Data: 00:6f:79:01:01
                ZigBee ZDO
                        Status = Success: (0x00)
                        NWK Address Of Interest: 0x796f
                        Match Length: 1
                        Match List
                                Endpoint 1: 0x01
**********************************************************************
```

Sniffer Capture 9.12: The ZigBee specification calls for a Unicast response and that's exactly what the Cirronet ZMN2400HP did.

Before we move on, I would like to show you something that is very important. Here's the IEEE 802.15.4 acknowledgment to the Match_Desc_resp message:

```
**********************************************************************
Frame 3 (Length = 5 bytes)
        Time Stamp: 10:20:21.008
        Frame Length: 5 bytes
        Capture Length: 5 bytes
        Link Quality Indication: 220
IEEE 802.15.4
        Frame Control: 0x0002
                .... .... .... .010  = Frame Type: Acknowledgment (0x0002)
                .... .... .... 0...  = Security Enabled: Disabled
```

```
            .... .... ...0 ....  = Frame Pending: No more data
            .... .... ..0. ....  = Acknowledgment Request: Acknowledgment
not required
            .... ... .0.. ....   = Intra PAN: Not within the PAN
            .... ..00 0... ....   = Reserved
            .... 00.. .... ....   = Destination Addressing Mode: PAN
identifier and address field are not present (0x0000)
            ..00 .... .... ....   = Reserved
            00.. .... .... ....   = Source Addressing Mode: PAN identifier
and address field are not present (0x0000)
        Sequence Number: 38
        Frame Check Sequence: Correct


0000:   02 00 26 .. ..                                      ..&..
*****************************************************************************
```

And, here's the APS application acknowledgment to the Match_Desc_resp message:

```
*****************************************************************************
Frame 4 (Length = 25 bytes)
        Time Stamp: 10:20:21.012
        Frame Length: 25 bytes
        Capture Length: 25 bytes
        Link Quality Indication: 220
IEEE 802.15.4
        Frame Control: 0x8861
            .... .... .... .001  = Frame Type: Data (0x0001)
            .... .... .... 0...   = Security Enabled: Disabled
            .... .... ...0 ....   = Frame Pending: No more data
            .... .... ..1. ....   = Acknowledgment Request: Acknowledgment
required
            .... .... .1.. ....   = Intra PAN: Within the PAN
            .... ..00 0... ....   = Reserved
            .... 10.. .... ....   = Destination Addressing Mode: Address
field contains a 16-bit short address (0x0002)
            ..00 .... .... ....   = Reserved
            10.. .... .... ....   = Source Addressing Mode: Address field
contains a 16-bit short address (0x0002)
        Sequence Number: 79
        Destination PAN Identifier: 0x307d
        Destination Address: 0x796f
        Source Address: 0x0000
        Frame Check Sequence: Correct
ZigBee NWK
        Frame Control: 0x0044
            .... .... .... ..00  = Frame Type: NWK Data (0x00)
            .... .... ..00 01..   = Protocol Version (0x01)
            .... .... 01.. ....   = Discover Route: Enable route discovery
(0x01)
            .... ...0 .... ....   = Reserved
            .... ..0. .... ....   = Security: Disabled
            0000 00.. .... ....   = Reserved
        Destination Address: 0x796f
```

```
        Source Address: 0x0000
        Radius = 7
        Sequence Number = 18
ZigBee APS
        Frame Control: 0x02
                .... ..10  = Frame Type: APS Ack (0x02)
                .... 00..  = Delivery Mode: Normal Unicast Delivery (0x00)
                ...0 ....  = Indirect Address Mode: Ignored
                ..0. ....  = Security: False
                .0.. ....  = Ack Request: Acknowledgment not required
                0... ....  = Reserved
        Destination Endpoint: 0x00
        Cluster Identifier: MatchDescRsp (0x0086)
        Profile Identifier: ZDP (0x0000)
        Source Endpoint: 0x00
```
****

This ZigBee stack stuff is kinda like algebra. What is done to one side must be done to the other. In order for the application to keep itself in sync with what's flowing through the ZigBee stack layers, checkpoints in the form of application and network acknowledgments must exist.

Now that we have obtained the lone remote device's network address, let's get greedy and see if we can assemble a message to assimilate its IEEE 802.15.4 address as well. A good beginning is the WinCom window and the data therein shown in Screen Capture 9.3.

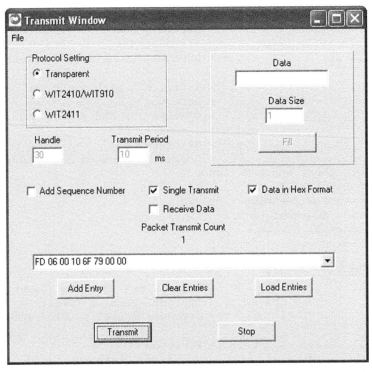

Screen Capture 9.3: The main ingredient of this mix is the network address we just obtained.

Again, cutting to the chase, the MSG Type is 0x10, Get IEEE Address. We are looking for the IEEE address of the network node addressed as 0x796F and only want the MAC address returned. There is no list to index. The Get IEEE Address argument list is notated in Figure 9.5.

Arguments (Get EEE Address)		
Network Address (2 bytes)	Request Type (1 byte)	Start Index (1 byte)

Figure 9.5: The Start Index value allows the retrieval of a number of devices at differing indexes into a list of devices that may be too long and overrun the limit of a single ZigBee packet size of 127 bytes.

A click on the Transmit button sends the packet in Sniffer Capture 9.13 along on its way.

Sniffer Capture 9.13
****
```
Frame 6 (Length = 32 bytes)
        Time Stamp: 10:23:09.381
        Frame Length: 32 bytes
        Capture Length: 32 bytes
        Link Quality Indication: 212
IEEE 802.15.4
        Frame Control: 0x8861
                .... .... .... .001 = Frame Type: Data (0x0001)
                .... .... .... 0... = Security Enabled: Disabled
                .... .... ...0 .... = Frame Pending: No more data
                .... .... ..1. .... = Acknowledgment Request: Acknowledgment
required
                .... .... .1.. .... = Intra PAN: Within the PAN
                .... ..00 0... .... = Reserved
                .... 10.. .... .... = Destination Addressing Mode: Address
field contains a 16-bit short address (0x0002)
                ..00 .... .... .... = Reserved
                10.. .... .... .... = Source Addressing Mode: Address field
contains a 16-bit short address (0x0002)
        Sequence Number: 80
        Destination PAN Identifier: 0x307d
        Destination Address: 0x796f
        Source Address: 0x0000
        Frame Check Sequence: Correct
ZigBee NWK
        Frame Control: 0x0004
                .... .... .... ..00 = Frame Type: NWK Data (0x00)
                .... .... ..00 01.. = Protocol Version (0x01)
                .... .... 00.. .... = Discover Route: Suppress route
discovery (0x00)
                .... ...0 .... .... = Reserved
                .... ..0. .... .... = Security: Disabled
                0000 00.. .... .... = Reserved
        Destination Address: 0x796f
```

```
            Source Address: 0x0000
            Radius = 7
            Sequence Number = 19
ZigBee APS
      Frame Control: 0x00
                  .... ..00  = Frame Type: APS Data (0x00)
                  .... 00..  = Delivery Mode: Normal Unicast Delivery (0x00)
                  ...0 ....  = Indirect Address Mode: Ignored
                  ..0. ....  = Security: False
                  .0.. ....  = Ack Request: Acknowledgment not required
                  0... ....  = Reserved
            Destination Endpoint: 0x00
            Cluster Identifier: IEEEAddrReq (0x0001)
            Profile Identifier: ZDP (0x0000)
            Source Endpoint: 0x00
ZigBee AF
      AF Header: 0x21
                  .... 0001  = Transaction Count: (0x01)
                  0010 ....  = Frame Type: MSG (0x02)
            Transaction 1
                  Transaction Sequence Number = 0x0d
                  ZigBee MSG
                        Transaction Length: 4
                        Transaction Data: 6f:79:00:00
                  ZigBee ZDO
                        NWK Address Of Interest: 0x796f
                        Request Type = Single Device Response: (0x00)
                        Start Index: 0

0000:   61 88 50 7d 30 6f 79 00 00 04 00 6f 79 00 00 07    a.P}0oy....oy...
0010:   13 00 00 01 00 00 00 21 0d 04 6f 79 00 00 .. ..    .......!..oy....
```
****

Sniffer Capture 9.13: See, this ZigBee stack stuff isn't so complicated when you can see what the layers are really doing.

The ZigBee APS and ZigBee AF sniff contents tell the whole story from the application point of view. The Cirronet ZigBee Coordinator application (our little Get IEEE Address message) wants the 64-bit IEEE 802.15.4 MAC address of the remote Cirronet node with a NWK address of 0x796F and nothing more. Here's what Cirronet node 0x796F returned:

****
```
ZigBee APS
      Frame Control: 0x40
                  .... ..00  = Frame Type: APS Data (0x00)
                  .... 00..  = Delivery Mode: Normal Unicast Delivery (0x00)
                  ...0 ....  = Indirect Address Mode: Ignored
                  ..0. ....  = Security: False
                  .1.. ....  = Ack Request: Acknowledgment required
                  0... ....  = Reserved
            Destination Endpoint: 0x00
            Cluster Identifier: IEEEAddrRsp (0x0081)
```

```
        Profile Identifier: ZDP (0x0000)
        Source Endpoint: 0x00
ZigBee AF
        AF Header: 0x21
                .... 0001  = Transaction Count: (0x01)
                0010 ....  = Frame Type: MSG (0x02)
        Transaction 1
                Transaction Sequence Number = 0x0d
                ZigBee MSG
                        Transaction Length: 13
                        Transaction Data: 00:ca:0a:00:00:03:66:30:00:6f:79:00:0
0
                ZigBee ZDO
                        Status = Success: (0x00)
                        IEEE Address of Remote Device: 0x0030660300000aca
                        Nework Address of Remote Device: 0x796f
                        Number of Associated Device: 0
                        Start Index: 0
```
****

I eliminated the IEEE 802.15.4 and NWK portions of the sniff as you're familiar with their contents. The data our little one-line application desires is embedded within the ZigBee AF portion of the Daintree Networks SNA sniff. In addition to the IEEE 802.15.4 acknowledgments, a final APS acknowledgment was issued, just as it was for the Match_Desc_req message.

About Cirronet

Headquartered in North Metro Atlanta, Cirronet has over 17 years of success in the development of innovative wireless products. Cirronet maintains regional sales offices throughout the U.S.

If you're interested in ZigBee, I can assure you that Cirronet is interested in you as my contact inside of Cirronet is the Vice President of Marketing, Tim Cutler. Tim didn't "delegate" me to one of his staff and was directly responsible for providing the Cirronet content for this book. Tim returns calls and answers emails even while he's doing that VP stuff. Thanks, Tim.

I'll bet Tim knows the band from Athens. Let's see, some of their hits include:

- Rock Lobster
- Love Shack

Still stumped? Fred and Cindy's band is named after a very famous military jet, the B52. Athens, Georgia, just happens to be the home of the University of Georgia. What a place to form the world's greatest party band.

What musical band of brothers, all hailing from the South, ended up in the North and in the musical history books?

CHAPTER **10**

Silicon Laboratories

In our look at the ZMD 900-MHz IEEE 802.15.4-compliant/ZigBee-ready transceiver, we saw what Silicon Laboratories brings to the IEEE 802.15.4 and ZigBee table as far as microcontrollers go. In this chapter, we'll take yet another look at yet another Silicon Laboratories microcontroller, the Silicon Laboratories C8051F121. This time, however, the Silicon Laboratories C8051F121 joins in with a Silicon Laboratories CP2101 USB-to-UART bridge and a Chipcon CC2420 transceiver to form a 2.4-GHz/ZigBee node. Photo 10.1 is a pixilation of a full-blown Silicon Laboratories 2.4-GHz ZigBee node.

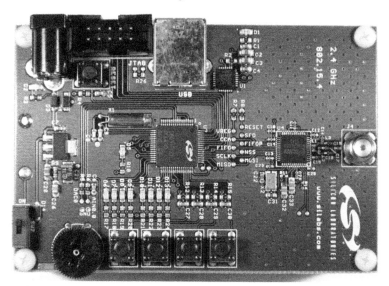

Photo 10.1: This is Silicon Laboratories' version of a ZigBee node. This design is very clean and easy to follow.

The Silicon Laboratories 2.4-GHz ZigBee node is built around the C8051F121. The C8051F121 is a high-performance 8051-derivative with mixed-signal capabilities. The C8051F121 has all of the attributes one would want in a high-end microcontroller with some extras. On-chip analog-to-digital converters have become commonplace. However, not many microcontrollers are equipped with a pair of on-chip DACs (digital-to-analog-converters). Temperature sensors aren't something you see as an on-chip peripheral either. The C8051F121, shown at low altitude in Photo 10.2, has both a pair of DACs and a temperature sensor embedded in its silicon.

Photo 10.2: The Silicon Laboratories 2.4-GHz ZigBee/IEEE 802.15.4 development board is very nicely laid out with the intention of teaching the user about the nuances of ZigBee and IEEE 802.15.4 hardware. Note the CC2420 signal trace tap point to the right of the C8051F121 in this shot.

There are no accessible RS-232 serial ports to be found on the Silicon Laboratories ZigBee development boards. Instead, the Silicon Laboratories engineers included a lone USB port, which is based on the Silicon Laboratories CP2101 USB-to-UART bridge. I got as close as I could to the CP2101 USB-to-UART bridge in Photo 10.3.

Photo 10.3: There isn't much you have to surround the CP2101 USB-to-UART bridge with to make it go. The USB port is used extensively with the personal computer-based demo packages that come with the Silicon Laboratories ZigBee development kit.

Can you find the discrete balun in Photo 10.4? Do you see it in Photo 10.5? To prove a point, the component layouts in both Photo 10.4 (Silicon Laboratories ZigBee development board radio) and Photo 10.5 (Texas Instruments/Chipcon's CC2420 ZigBee radio) are strikingly similar for good reason. Why try to beat up and improve something that the original manufacturer has perfected and told you to use?

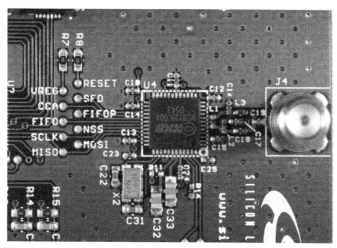

Photo 10.4: Here's a low-level shot of the Silicon Laboratories CC2420 radio layout.

Photo 10.5: Here's a shot of the original manufacturer's CC2420 radio layout. Texas Instruments/Chipcon advises users of the CC2420 to use their RF, layout as it has been extensively tested and proven.

The C8051F121's peripheral set is nice, but we've got ZigBee things to do. Once we get on the air, then the C8051F121's on-chip mixed-signal peripherals could be put to use to provide the data that will ride in the ZigBee packets.

The concept of primitives is conveyed in the IEEE 802.15.4 specification. However, you and I have already found that the implementation of the primitive principle can differ from manufacturer to manufacturer. The Silicon Laboratories ZigBee node in Photo 10.1 was supplied with a library package specifically targeted at the NWK layer of the ZigBee stack called the Silicon Laboratories ZigBee Network Layer Interface. The Silicon Laboratories ZigBee Network Layer Interface utilizes function calls and a globally shared buffer scheme to implement primitives.

Messages flowing between the application and network layers use function calls or a shared buffer as transport. Primitive traffic originating at the application layer that is aimed at the NWK layer rides on function calls. Network traffic flowing upstream to the application layer uses indication primitives, which are stored in the shared buffer. The application layer must interrogate the shared buffer to check for incoming traffic from the network layer.

Depending upon the type of confirmation primitive, there may be one or multiple pieces of data riding within the primitive's data structure. For confirmation primitives with a single parameter, which is normally a status indicator, the confirmation primitive's parameter is passed as a return value of the function call. If a confirmation primitive contains multiple parameters, the function will store the primitive's parameters in the shared buffer. It is then up to the requester to check for and retrieve the confirmation parameters. In the Silicon Laboratories ZigBee Network Layer Interface framework, all primitive requests are initiated with a function call. Now that you have been familiarized with the Silicon Laboratories ZigBee Network Layer Interface rules and regulations, let's walk through converting an NWK-generated message into RF particles.

Before we begin our mission to send a message from the NWK layer, I must bring to your attention that Silicon Laboratories also provides a similar library of functions for the MAC layer called the 802.15.4 MAC Layer Application Programming Interface. As you have quickly deduced, the 802.15.4 MAC API is aimed at the MLME and MCPS MAC primitives. The Silicon Laboratories ZigBee Network Layer Interface library contains the MAC API functionality. However, the Silicon Laboratories NWK library is provided in a package that does not allow the programmer direct access to the MAC primitives.

The Silicon Laboratories ZigBee NWK library contains a series of system initialization function calls that are necessary to execute before embarking on the use of the ZigBee NWK library resources in an application. Here is a list of the system initialization calls in the correct run order:

- DISABLE_GLOBAL_INT()   Disable Global Interrupts
- SystemInit ()   Initialize the System Hardware
- CC2420Init()   Initialize the Transceiver
- EINT_Init()   Initialize the Transceiver Interrupt
- MAC_Init()   Initialize the MAC

- macInitEnv()   Set Default PIB settings
- mlmeResetRequest(FALSE)   Reset the mac
- netInit()   Initialize the NWK layer
- ENABLE_GLOBAL_INT()   Enable Global Interrupts

The Silicon Laboratories ZigBee NWK library combines all of the system initialization steps into a single function call, InitAllSystem. Unfortunately, the source code for the Silicon Laboratories library is not something that I have access to. So, we won't be breaking down the system initialization code as walking through 8051 assembler won't be any fun for you or me.

If you've trusted me (and the ZigBee specification and the IEEE 802.15.4 specification) thus far, you know that we can apply the Silicon Laboratories ZigBee NWK library to a PAN Coordinator, a ZigBee Router and a ZigBee End Device. So, let's synchronize at this moment and declare the Silicon Laboratories ZigBee PAN Coordinator powered up and initialized.

In this scenario, you and I are the application as we are sending commands to the NWK layer from above. We've got a brand new shiny Silicon Laboratories ZigBee module that we need to do something with. However, we can't do anything ZigBee without first establishing a PAN. But, before we can form a PAN we must assign a PAN Coordinator. Our Silicon Laboratories ZigBee hardware has the punch needed to be a PAN Coordinator. So, we'll issue a Silicon Laboratories ZigBee NWK function call to crown our Silicon Laboratories ZigBee hardware king of a PAN. The network formation function, which represents the NLME-NET-WORK-FORMATION primitive, looks like this:

```
************************************************************************
nlmeNetworkFormationRequest(0x00000800, 5, 15, 15, 0x0002, FALSE)
************************************************************************
```

The prerequisites for issuing the nlmeNetworkFormationRequest function call are that the target device be an FFD that is not already engaged with another network and that NLME-RESET be issued prior to the nlmeNetworkFormationRequest function call. Both of the prerequisites have been fulfilled as NLME-RESET was issued in the netInit system initialization function and our target Silicon Laboratories C8051F121-based ZigBee hardware is more than capable of taking on a ZigBee PAN Coordinator role.

In the nlmeNetworkFormationRequest function's arguments I have limited the channel scan to channel 11 and declared a nonBeacon network with a PAN ID of 0x0002. As long as I didn't go hog wild on the hold time, scan duration here is irrelevant as I've effectively forbidden a multiple channel scan by signaling out channel 11. In the true fashion of ZigBee Coordinators, the Silicon Laboratories ZigBee Coordinator hardware needs no battery-life extension support and I've reflected that with a FALSE in the battery life extension argument.

Following the execution of the nlmeNetworkFormationRequest function, the Daintree Networks SNA posted the expected Beacon Request frame followed by a totally unexpected Coordinator Realignment frame, which I've posted for you in Sniffer Capture 10.1. After some careful investigation of the Silicon Laboratories 802.15.4 library documentation, I found that the coordRealignment bit within the MLME-START primitive's data structure was set to TRUE. The reasoning behind this is to provide a person using a Sniffer (that's us) a positive visual notification that the new PAN has indeed been created. Hopefully, you're not wondering how an MLME-START process got into the NWK mix. Remember, everything flows downhill (top to bottom) in the ZigBee stack when the application kicks off a request and each layer performs its part of the job before sending the data out of the RF pipe or providing the stuff that the application requested.

Sniffer Capture 10.1

```
**************************************************************************
Frame 2 (Length = 27 bytes)
        Time Stamp: 15:53:00.509
        Frame Length: 27 bytes
        Capture Length: 27 bytes
        Link Quality Indication: 208
IEEE 802.15.4
        Frame Control: 0xc803
                .... .... .... .011  = Frame Type: Command (0x0003)
                .... .... .... 0...  = Security Enabled: Disabled
                .... .... ...0 ....  = Frame Pending: No more data
                .... .... ..0. ....  = Acknowledgment Request: Acknowledgment
not required
                .... .... .0.. ....  = Intra PAN: Not within the PAN
                .... ..00 0... ....  = Reserved
                .... 10.. .... ....  = Destination Addressing Mode: Address
field contains a 16-bit short address (0x0002)
                ..00 .... .... ....  = Reserved
                11.. .... .... ....  = Source Addressing Mode: Address field
contains a 64-bit extended address (0x0003)
        Sequence Number: 13
        Destination PAN Identifier: 0xffff
        Destination Address: 0xffff
        Source PAN Identifier: 0x0002
        Source Address: 0x000b5700000009c6
        MAC Payload
                Command Frame Identifier = Coordinator Realignment: (0x08)
                PAN Identifier: 0x0002
                Coordinator Short Address: 0x0000
                Logical Channel: 11
                Short Address: 0x0000
        Frame Check Sequence: Correct

0000:   03 c8 0d ff ff ff ff 02 00 c6 09 00 00 00 57 0b    .Hd...........W.
0010:   00 08 02 00 00 00 0b 00 00 .. ..                   ..........
**************************************************************************
```

Sniffer Capture 10.1: This is an optional frame to fire off following the creation of a ZigBee PAN. It's a good idea in this sense as it totally identifies the new PAN and its Coordinator to the person observing the Sniffer capture (that be you).

Now that the Silicon Laboratories MAC code has made sure that we know a valid ZigBee PAN was created, let's have a new PAN party and "associate" with some friends. Before we can party we have to send some invitations. Our ZigBee PAN party invites all look alike and are based on the NLME-PERMIT-JOINING primitive. The final invitation looks like this:

```
*************************************************************************
      nlmePermitJoiningRequest(0xFF)
*************************************************************************
```

The 0xFF indicates that the ZigBee PAN Coordinator is open for accepting requests to join the PAN as long as there's room for new guests. And, the invites are never really sent out as the nlmePermitJoiningRequest function is a signal to set internal flags in upper layers to allow the MAC to set a flag to allow nodes to join the PAN.

At this point, let's turn on one of those "guests" and start our ZigBee PAN party. We'll use another Silicon Laboratories ZigBee module that is identical in build to our Silicon Laboratories ZigBee module that is currently acting as the ZigBee PAN Coordinator. However, we will pretend (not assume) that the Silicon Laboratories ZigBee module we're bringing up cannot function as an FFD, which means it will be relegated to ZigBee End Device service. All of the things we did to initialize the Silicon Laboratories ZigBee Coordinator are also performed within the C8051F121 on the up-and-coming ZigBee End Device. However, instead of starting a ZigBee PAN, which it can't do as an RFD, the newly initialized Silicon Laboratories ZigBee hardware executes the following function:

```
*************************************************************************
      nlmeNetworkDiscoveryRequest(0x00000800, 5)
*************************************************************************
```

The nlmeNetworkDiscoveryRequest function is derived from the NLME-NETWORK-DIS-COVERY primitive. We have just told the not-yet-associated ZigBee End Device to search the sky for networks. Instead of forcing the soon-to-be ZigBee End Device to search all of the available 2.4-GHz channels, I forced the scan to channel 11 only. A confirmation (NLME-NETWORK-DISCOVERY.confirm in primitive speak) was returned to the shared buffer, which I just happened to have an eye on in Screen Capture 10.1.

The first byte in the nlmeconfirm buffer is the NetworkCount, whose value represents how many network descriptor structures the buffer is holding. All of the bytes that follow will fall nicely into the slots of the NETWORK_DESCRIPTOR structure in Code Snippet 10.1. A search status byte follows the securityLevel byte in the dump shown in Screen Capture 10.1. What you don't see in Code Snippet 10.1 is the passing of the confirmID byte that is used as a signal and descriptor to let other layers know that the data is related to a network discovery operation.

Name	Value
⊟ `nlmeconfirm.buffer,0x10	X:0x000C8B [...]
[0]	0x01
[1]	0x00
[2]	0x02
[3]	0x0B
[4]	0x0F
[5]	0x00
[6]	0x0F
[7]	0x0F
[8]	0x01
[9]	0x07
[10]	0x00
[11]	0x00
[12]	0x00

Locals \ Watch #1 \ **Watch #2** \ Call Stack

Screen Capture 10.1: I stopped the C8051F121's execution when the nlmeNetworkDiscoveryRequest function completed to get this view of the globally shared nlmeconfirm buffer.

Code Snippet 10.1
****

```
typedef struct{
BYTE NetworkCount;
NETWORK_DESCRIPTOR nwkDescriptor[MAX_USE_CHANNEL_COUNT];
MAC_ENUM Status;
}NLME_NETWORK_DISCOVERY_CONFIRM;

typedef struct {
WORD panID;
BYTE logicalChannel;
BYTE stackProfile;
BYTE zigBeeVersion;
BYTE BeaconOrder;
BYTE superFrameOrder;
BOOL permitJoining;
BYTE securityLevel;
}NETWORK_DESCRIPTOR;
```
****

Code Snippet 10.1: This is what you don't see in the Daintree Networks SNA traces because this kind of stuff is going on between the layers and is not transmitted. This represents the Silicon Laboratories ZigBee NWK library's use of the shared buffer primitive passing technique.

A search status byte that is equal to 0x00 (SUCCESS) means that at least one network was discovered. That's good because there is only one network in operation, ours. Now that a functioning ZigBee PAN has been located, the ZigBee End Device can request to join the ZigBee PAN party by issuing this function call:

****

```
nlmeJoinRequest(0x0002, FALSE, FALSE, 0x00000800, 5, 1, TRUE, FALSE);
```
****

The nlmeJoinRequest function is a derivative of the NLME-JOIN primitive. I'm sure you can figure out most of what the nlmeJoinRequest function call is doing by the PANID and channel arguments. Basically, we're asking to join a PAN IDed as 0x0002 on channel 11 as a ZigBee End Device using a mains power supply because we will leave the receiver on when the ZigBee node is idle. And, by the way, hold the security. Here's what goes down:

(Note that I will not post the acknowledgment sniffs in this thread of events. I will, however, make you aware of their locations in the thread.)

The wanna-be ZigBee End Device actively scans channel 11 by issuing a Beacon Request frame.

```
*****************************************************************************
Frame 3 (Length = 10 bytes)
        Time Stamp: 09:29:17.288
        Frame Length: 10 bytes
        Capture Length: 10 bytes
        Link Quality Indication: 152
IEEE 802.15.4
        Frame Control: 0x0803
                .... .... .... .011 = Frame Type: Command (0x0003)
                .... .... .... 0... = Security Enabled: Disabled
                .... .... ...0 .... = Frame Pending: No more data
                .... .... ..0. .... = Acknowledgment Request: Acknowledgment
not required
                .... .... .0.. .... = Intra PAN: Not within the PAN
                .... ..00 0... .... = Reserved
                .... 10.. .... .... = Destination Addressing Mode: Address
field contains a 16-bit short address (0x0002)
                ..00 .... .... .... = Reserved
                00.. .... .... .... = Source Addressing Mode: PAN identifier
and address field are not present (0x0000)
        Sequence Number: 167
        Destination PAN Identifier: 0xffff
        Destination Address: 0xffff
        MAC Payload
                Command Frame Identifier = Beacon Request: (0x07)
        Frame Check Sequence: Correct

0000:   03 08 a7 ff ff ff ff 07 .. ..                      ..'.......
*****************************************************************************
```

The ZigBee PAN Coordinator (ID = 0x0002) replies by transmitting an informational Beacon.

```
*****************************************************************************
Frame 4 (Length = 19 bytes)
        Time Stamp: 09:29:17.294
        Frame Length: 19 bytes
        Capture Length: 19 bytes
        Link Quality Indication: 212
IEEE 802.15.4
        Frame Control: 0x8000
```

```
                      .... .... .... .000  = Frame Type: Beacon (0x0000)
                      .... .... .... 0...  = Security Enabled: Disabled
                      .... .... ...0 ....  = Frame Pending: No more data
                      .... .... ..0. ....  = Acknowledgment Request: Acknowledgment
not required
                      .... .... .0.. ....  = Intra PAN: Not within the PAN
                      .... ..00 0... ....  = Reserved
                      .... 00.. .... ....  = Destination Addressing Mode: PAN
identifier and address field are not present (0x0000)
                      ..00 .... .... ....  = Reserved
                      10.. .... .... ....  = Source Addressing Mode: Address field
contains a 16-bit short address (0x0002)
        Sequence Number: 177
        Source PAN Identifier: 0x0002
        Source Address: 0x0000
        MAC Payload
              Superframe Specification: 0xcfff
                      .... .... .... 1111  = Beacon Order (0x000f)
                      .... .... 1111 ....  = Superframe Order (0x000f)
                      .... 1111 .... ....  = Final CAP Slot (0x000f)
                      ...0 .... .... ....  = Battery Life Extension: Disabled
                      ..0. .... .... ....  = Reserved
                      .1.. .... .... ....  = PAN Coordinator: Transmitter is
a PAN Coordinator
                      1... .... .... ....  = Association Permit: Coordinator
accepting Association Requests
              GTS Specification: 0x80
                      .... .000  = GTS Descriptor Count (0x00)
                      .000 0...  = Reserved
                      1... ....  = GTS Permit: Coordinator accepting GTS
Requests
              Pending Address Specification: 0x00
                      .... .000  = Number of short Addresses pending: 0
                      .... 0...  = Reserved
                      .000 ....  = Number of extended Addresses pending: 0
                      0... ....  = Reserved
              Beacon Payload
              MAC Payload: 00:0f:0f:00:00:00
        Frame Check Sequence: Correct

0000:  00 80 b1 02 00 00 00 ff cf 80 00 00 0f 0f 00 00    ..1.....O.......
0010:  00 .. ..
***********************************************************************
```

The wanna-be ZigBee End Device issues an Association Request.

```
***********************************************************************
Frame 5 (Length = 21 bytes)
        Time Stamp: 09:29:28.836
        Frame Length: 21 bytes
        Capture Length: 21 bytes
        Link Quality Indication: 156
IEEE 802.15.4
```

```
        Frame Control: 0xc823
                .... .... .... .011  = Frame Type: Command (0x0003)
                .... .... .... 0...  = Security Enabled: Disabled
                .... .... ...0 ....  = Frame Pending: No more data
                .... .... ..1. ....  = Acknowledgment Request: Acknowledgment
required
                .... .... .0.. ....  = Intra PAN: Not within the PAN
                .... ..00 0... ....  = Reserved
                .... 10.. .... ....  = Destination Addressing Mode: Address
field contains a 16-bit short address (0x0002)
                ..00 .... .... ....  = Reserved
                11.. .... .... ....  = Source Addressing Mode: Address field
contains a 64-bit extended address (0x0003)
        Sequence Number: 168
        Destination PAN Identifier: 0x0002
        Destination Address: 0x0000
        Source PAN Identifier: 0xffff
        Source Address: 0x000b570000000987
        MAC Payload
                Command Frame Identifier = Association Request: (0x01)
                Capability Information: 0x8c
                        .... ...0  = Alternate PAN Coordinator: Not capable of
becoming PAN Coordinator
                        .... ..0.  = Device Type: RFD
                        .... .1..  = Power Source: Receiving power from
alternating current mains
                        .... 1...  = Receiver on when idle: Enables receiver
when idle
                        ..00 ....  = Reserved
                        .0.. ....  = Security Capability: Not capable of using
security suite
                        1... ....  = Allocate Address: Coordinator should
allocate short address
        Frame Check Sequence: Correct

0000:   23 c8 a8 02 00 00 00 ff ff 87 09 00 00 00 57 0b    #H(..........W.
0010:   00 01 8c .. ..                                     .....
```

****

The ZigBee PAN Coordinator transmits the required acknowledgment (Frame 6) and the wanna-be ZigBee End Device transmits a Data Request.

****

```
Frame 7 (Length = 18 bytes)
        Time Stamp: 09:29:29.332
        Frame Length: 18 bytes
        Capture Length: 18 bytes
        Link Quality Indication: 156
IEEE 802.15.4
        Frame Control: 0xc863
                .... .... .... .011  = Frame Type: Command (0x0003)
                .... .... .... 0...  = Security Enabled: Disabled
```

```
                    .... .... ...0 ....  = Frame Pending: No more data
                    .... .... ..1. ....  = Acknowledgment Request: Acknowledgment
required
                    .... .... .1.. ....  = Intra PAN: Within the PAN
                    .... ..00 0... ....  = Reserved
                    .... 10.. .... ....  = Destination Addressing Mode: Address
field contains a 16-bit short address (0x0002)
                    ..00 .... .... ....  = Reserved
                    11.. .... .... ....  = Source Addressing Mode: Address field
contains a 64-bit extended address (0x0003)
        Sequence Number: 169
        Destination PAN Identifier: 0x0002
        Destination Address: 0x0000
        Source Address: 0x000b570000000987
        MAC Payload
                Command Frame Identifier = Data Request: (0x04)
        Frame Check Sequence: Correct


0000:    63 c8 a9 02 00 00 00 87 09 00 00 00 57 0b 00 04    cH).........W...
0010:    .. ..                                                                ..
```

****

Again, the ZigBee PAN Coordinator returns an acknowledgment (Frame 8) as requested and follows the acknowledgment up with an Association Response.

****

```
Frame 9 (Length = 27 bytes)
        Time Stamp: 09:29:29.337
        Frame Length: 27 bytes
        Capture Length: 27 bytes
        Link Quality Indication: 212
IEEE 802.15.4
        Frame Control: 0xcc63
                    .... .... .... .011  = Frame Type: Command (0x0003)
                    .... .... .... 0...  = Security Enabled: Disabled
                    .... .... ...0 ....  = Frame Pending: No more data
                    .... .... ..1. ....  = Acknowledgment Request: Acknowledgment
required
                    .... .... .1.. ....  = Intra PAN: Within the PAN
                    .... ..00 0... ....  = Reserved
                    .... 11.. .... ....  = Destination Addressing Mode: Address
field contains a 64-bit extended address (0x0003)
                    ..00 .... .... ....  = Reserved
                    11.. .... .... ....  = Source Addressing Mode: Address field
contains a 64-bit extended address (0x0003)
        Sequence Number: 254
        Destination PAN Identifier: 0x0002
        Destination Address: 0x000b570000000987
        Source Address: 0x000b5700000009c6
        MAC Payload
                Command Frame Identifier = Association Response: (0x02)
                Short Address: 0x155a
```

```
           Association Status: Association Successful (0x00)
        Frame Check Sequence: Correct

0000:    63 cc fe 02 00 87 09 00 00 00 57 0b 00 c6 09 00    cL~.......W..F..
0010:    00 00 57 0b 00 02 5a 15 00 .. ..                   ..W...Z....
*********************************************************************************
```

The wanna-be ZigBee End Device has been accepted (Association Status: Association Successful (0x00)) into the ZigBee PAN and transmits the requested acknowledgment frame (Frame 10).

When it all goes down as planned, you can put 0x00, 0x02 and 0x00 in the first three bytes of the nlmeconfirm buffer dump in Screen Capture 10.1 and consider the Silicon Laboratories ZigBee End Device node as part of the ZigBee PAN party.

The next thing a new ZigBee End Device might want to do is send along some data to its new boss. Since we're pretending here, let's put some data in the payload that you can readily pick out of the Daintree Networks SNA sniff. I've coded up a function in Code Snippet 10.2 that will send my name from the ZigBee End Device to the ZigBee PAN Coordinator.

Code Snippet 10.2

```
*********************************************************************************
void SendData(void)
{
    NLDE_DATA_REQUEST        xdata       NldeDataRequest;
    unsigned char            xdata       dataLen;

    DataBuffer[0] = 'F';
    DataBuffer[1] = 'R';
    DataBuffer[2] = 'E';
    DataBuffer[3] = 'D';
    dataLen = 0x04;

    NldeDataRequest.dstAddr           = 0x0000;
    NldeDataRequest.nsduLength        = dataLen;
    NldeDataRequest.pNsdu             = DataBuffer;
    NldeDataRequest.nsduHandle        = 1;
    NldeDataRequest.broadcastRadius   = 1;
    NldeDataRequest.discoverRoute     = FALSE;
    NldeDataRequest.securityEnable    = FALSE;

    nldeDataRequest((NLDE_DATA_REQUEST*)&(NldeDataRequest.dstAddr));
}
*********************************************************************************
```

Code Snippet 10.2: This is the Silicon Laboratories function version of passing an NLDE-DATA.request primitive. The presence of the data request to the layers that need to process the primitive data in the shared buffer is indicated by a confirmID value of N_DATA_IND.

Here's what happened when I executed the SendData function:

```
*********************************************************************************
Frame 11 (Length = 23 bytes)
        Time Stamp: 10:05:19.389
```

```
            Frame Length: 23 bytes
            Capture Length: 23 bytes
            Link Quality Indication: 128
IEEE 802.15.4
        Frame Control: 0xaa61
                .... .... .... .001  = Frame Type: Data (0x0001)
                .... .... .... 0...  = Security Enabled: Disabled
                .... .... ...0 ....  = Frame Pending: No more data
                .... .... ..1. ....  = Acknowledgment Request: Acknowledgment
required
                .... .... .1.. ....  = Intra PAN: Within the PAN
                .... ..10 0... ....  = Reserved
                .... 10.. .... ....  = Destination Addressing Mode: Address
field contains a 16-bit short address (0x0002)
                ..00 .... .... ....  = Reserved
                10.. .... .... ....  = Source Addressing Mode: Address field
contains a 16-bit short address (0x0002)
        Sequence Number: 99
        Destination PAN Identifier: 0x0002
        Destination Address: 0x0000
        Source Address: 0x155a
        Frame Check Sequence: Correct
ZigBee NWK
        Frame Control: 0x0004
                .... .... .... ..00  = Frame Type: NWK Data (0x00)
                .... .... ..00 01..  = Protocol Version (0x01)
                .... .... 00.. ....  = Discover Route: Suppress route
discovery (0x00)
                .... ...0 .... ....  = Reserved
                .... ..0. .... ....  = Security: Disabled
                0000 00.. .... ....  = Reserved
        Destination Address: 0x0000
        Source Address: 0x155a
        Radius = 1
        Sequence Number = 113
NWK Payload: 46:52:45:44

0000:   61 aa 63 02 00 00 00 5a 15 04 00 00 00 5a 15 01    a*c....Z.....Z..
0010:   71 46 52 45 44 .. ..                               qFRED..
********************************************************************************
```

This is a good time to see how our wanna-be ZigBee End Device, now a full-fledged ZigBee End Device, gracefully leaves the ZigBee PAN party. The formal way to exit the ZigBee PAN party is to call upon the NLME-LEAVE.request primitive in this Silicon Laboratories function form:

```
********************************************************************************
    nlmeLeaveRequest(deviceAddress);

    where deviceAddress = 0x000B570000000987
********************************************************************************
```

The deviceAddress happens to be our ZigBee End Device's 64-bit IEEE address. The confirmID that is passed in the primitive indication is N_LEAVE_IND. Here's what happened when the ZigBee End Device said goodbye to the ZigBee PAN party:

```
************************************************************************
Frame 11 (Length = 21 bytes)
        Time Stamp: 14:31:42.266
        Frame Length: 21 bytes
        Capture Length: 21 bytes
        Link Quality Indication: 112
IEEE 802.15.4
        Frame Control: 0xaa41
                .... .... .... .001  = Frame Type: Data (0x0001)
                .... .... .... 0...  = Security Enabled: Disabled
                .... .... ...0 ....  = Frame Pending: No more data
                .... .... ..0. ....  = Acknowledgment Request: Acknowledgment
not required
                .... .... .1.. ....  = Intra PAN: Within the PAN
                .... ..10 0... ....  = Reserved
                .... 10.. .... ....  = Destination Addressing Mode: Address
field contains a 16-bit short address (0x0002)
                ..00 .... .... ....  = Reserved
                10.. .... .... ....  = Source Addressing Mode: Address field
contains a 16-bit short address (0x0002)
        Sequence Number: 75
        Destination PAN Identifier: 0x0002
        Destination Address: 0xffff
        Source Address: 0x155a
        Frame Check Sequence: Correct
ZigBee NWK
        Frame Control: 0x00c5
                .... .... .... ..01  = Frame Type: NWK Command (0x01)
                .... .... ..00 01..  = Protocol Version (0x01)
                .... .... 11.. ....  = Discover Route: Reserved (0x03)
                .... ...0 .... ....  = Reserved
                .... ..0. .... ....  = Security: Disabled
                0000 00.. .... ....  = Reserved
        Destination Address: 0xffff
        Source Address: 0x155a
        Radius = 0
        Sequence Number = 60
        NWK Payload
                NWK Command Identifier = Leave: (0x04)
                Options: 0xb0
                        ...1 0000  = Reserved
                        ..1. ....  = Rejoin: True
                        .0.. ....  = Request/Indication: The device plans to
leave the network (0x0)
                        1... ....  = Remove Children: Children of the device
will be removed (0x1)

0000:   41 aa 4b 02 00 ff ff 5a 15 c5 00 ff ff 5a 15 00     A*K....Z.E...Z..
0010:   3c 04 b0 .. ..                                      <.0..
************************************************************************
```

You're very good at reading Daintree Networks SNA traces by now and there's not much more that needs to be said.

OK…You're spec-ed up on NWK operations, thanks to the folks at Silicon Laboratories. I must say that I thoroughly enjoyed working with the Silicon Laboratories development tools. I blended the Silicon Laboratories USB Debug Adapter (Photo 10.6) with a full version of Keil's uVision3 to produce the RF you read through in this chapter.

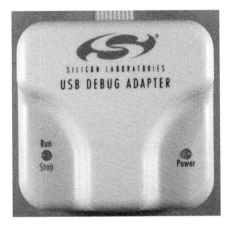

Photo 10.6: Remember this? I've worked with lots of debug devices in my time and this unassuming little box rates in the Top 10 for its ease of use and lack of unnecessary complexity.

If you're an 8051 type, working with the C8051F121 and uVision3 will be a pleasant surprise, as I was able to flow effortlessly between the uVision3 debug and edit modes.

The Silicon Laboratories development kit I used in this chapter comes with a total of six C8051F121-based ZigBee development boards. Just for grins, I loaded another one of the Silicon Laboratories ZigBee development boards as a ZigBee End Device and fired it up. Screen Capture 10.2 gives you an idea of the fun I was having while piloting Daintree Networks SNA and the Silicon Laboratories ZigBee modules.

About Silicon Laboratories

Silicon Laboratories is based in Austin, Texas, and designs and develops analog-intensive, mixed-signal integrated circuits. Although Silicon Laboratories does lots of things well, their 8051 products are among the best you will find anywhere.

The Elgins, under the watchful eye of Berry Gordy, became The Temptations. Can you name all of the "Classic 5" Temps and which song vaulted them into stardom?

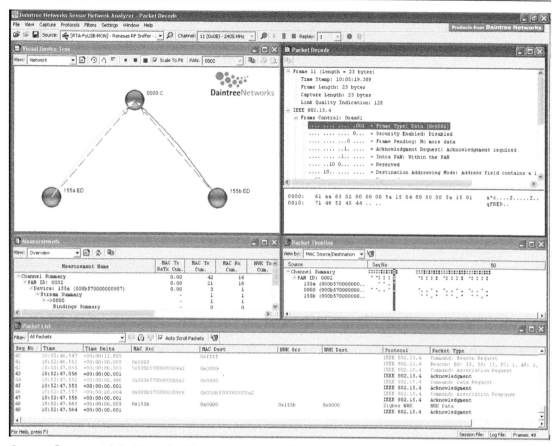

Screen Capture 10.2: Note the ZigBee PAN Coordinator and its two ZigBee End Device moons in the Visual Device Tree window. Daintree Networks SNA is a wonderful ZigBee tool.

CHAPTER **11**

Renesas

From the time you sifted through the very first Daintree Networks SNA sniff, you have been experiencing the power of the Renesas M16C series of 16-bit microcontrollers. A Renesas M30280FA sailing with a Texas Instruments/Chipcon CC2420 IEEE 802.15.4-compliant/Zig-Bee-ready transceiver has been (and still is) performing the hardware portion of the Daintree Networks SNA Sniffer duty.

The Renesas M30280FA is a formidable 16-bit microcontroller. With 8K of on-chip SRAM and 96K of program Flash, the M30280FA is a very good choice for commanding ZigBee and IEEE 802.15.4 nodes. There is ample general-purpose I/O (up to 71 I/O lines) available to drive the ZigBee radio and perform ZigBee sensor shepherding. In addition to easily handling IEEE 802.15.4 and ZigBee chores, the M30280FA has the ability to speak via a number of UART ports, SPI channels and even I²C. I put the macro lens on one of the Renesas ZigBee development boards in Photo 11.1.

Photo 11.1: This unassuming microcontroller is really a 16-bit monster on the loose. The M30280FA is shown here in its command and control position along with its first officer and first mate.

There are currently two variants of the Renesas M30280FA-based ZigBee development boards. The only difference in the boards is the frequency and manufacturer of the IEEE 802.15.4-compliant/ZigBee transceivers. Photo 11.2 is a macro-lens view of the 900-MHz Renesas M30280FA variant. There should be something very familiar about the transceiver IC in Photo 11.2.

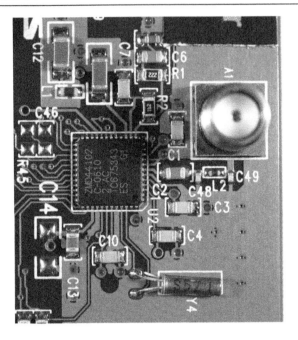

Photo 11.2: The 900-MHz variant of the Renesas M30280FA-
based ZigBee/IEEE 802.15.4
development board is powered
on the RF side by a ZMD44102
900-MHz IEEE 802.15.4-compli-
ant/ZigBee-ready transceiver IC.

If you didn't get here on the chapter-hopping red-eye flight, you're already spec-ed up on the inner workings of the ZMD44102 900-MHz IEEE 802.15.4-compliant/ZigBee-ready transceiver. And, if you didn't sleepwalk through the Texas Instruments/Chipcon chapter, the IEEE 802.15.4-compliant/ZigBee-ready transceiver IC from Texas Instruments/Chipcon in Photo 11.3 should elicit some sentimental thoughts as well.

Photo 11.3: This implementation of the CC2420 differs ever
so slightly from the Texas Instru-
ments/Chipcon CC2420 layout
template. I ran this Renesas
M30280FA configuration for
24 hours per day for months
with absolutely no problems or
failures.

Now that you've been introduced to all of the Renesas M30280FA ZigBee development-board players, a composite shot of the Renesas M30280FA-based ZigBee/IEEE 802.15.4 development board lies in Photo 11.4.

Photo 11.4: The Renesas M30280FA-based ZigBee development board also includes on-board light and thermal sensors that are attached to the M30280FA's analog-to-digital converter module.

The Renesas M30280FA-based ZigBee development board is supported by a pair of dongles, both of which connect to the M30280FA-based ZigBee development board via the 10-pin male header you see in the upper left quadrant of Photo 11.4. For packet-sniffing purposes, the Renesas M30280FA development board is loaded with Renesas packet-sniffer firmware that interfaces via the Sniffer dongle to the Daintree Networks SNA application running on a personal computer. The Renesas packet Sniffer dongle is shown in Photo 11.5.

Support for firmware loading and debugging is provided by a similar-looking dongle. Yes, I did take them apart. The electronics inside the case in Photo 11.5 are identical to the electronics buried inside of the plastic in Photo 11.6. The RF Sniffer Interface label was plastered over the original silkscreen you see in Photo 11.6. Only the firmware running on a Renesas Mitsubishi-era microcontroller differentiates the two dongles.

The RTA-FoUSB-MON dongle is used by both the debugging side of the Renesas High-performance Embedded Workshop (HEW) IDE and the Flash Over USB (FoUSB) programming application, which I've captured as a screen shot in Screen Capture 11.1.

Photo 11.5: What you don't see is a mini-USB connector on the left panel of this little device. When things are popping around the PAN, the LEDs in the recess are flickering like mad.

Photo 11.6: The names have been changed to protect the innards.

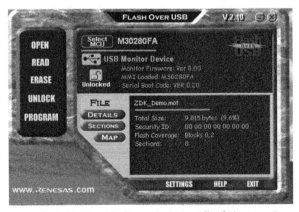

Screen Capture 11.1: The FoUSB window gives us all of the pertinent information about the file that is loaded in its buffer.

FoUSB is a necessary component of the M30280FA ZigBee development system, as it is used to download the Renesas ZigBee stack and RTOS to the M30280FA. Once the stack and RTOS are successfully installed in the M30280FA program Flash, we can use the Renesas ZigBee API to call upon the ZigBee stack's functionality.

The Renesas ZigBee stack is comprised of everything except the user application. Every other layer of the ZigBee stack is realized within the Renesas ZigBee stack code. All we have to do is access the services of the Renesas ZigBee stack using the Renesas ZigBee API calls. As we have seen in our ZigBee travels, everyone has a spin on the way they execute the laws of the ZigBee specification. To keep the application from having to concern itself with the relationship between request primitives and confirm primitives, the Renesas application-layer interface performs the relationship check for the programmer by providing a blocking API function to the upper layer. In other words, instead of the asynchronous primitive interface called out in the ZigBee specification, the Renesas application-layer interface provides a synchronous lock-step primitive interface. Despite this minor difference in operation, the Renesas API calls listed for you in Table 11.1 correspond to a related ZigBee specified primitive in name and parameter content. The Renesas API calls are relatively few in number but are all useful tools in real-world ZigBee applications. All of the Renesas API calls in Table 11.1 are NWK related.

Type	Function Name	Description
Network primitive related API function	ZbReset	Initialization
	ZbDiscoverNetwork	Discover network
	ZbFormNetwork	Form network
	ZbStartRouter	Start router
	ZbPermitJoining	Permit joining
	ZbJoin	Join to network
	ZbDirectJoin	Direct join
	ZbLeave	Leave from Network
	ZbDataSnd	Send data
	ZbSync	Polling data
	ZbSet	Set attribute
	ZbGet	Refer attribute
Callback function	ZbDataIndication	Data receive indication
	ZbJoinIndication	Child join indication
	ZbLeaveIndication	Leave indication

Type	Function Name	Description
Other API function	ZbMacAddr_Get	Get MAC address
	ZbShortAddr_Get	Get short address

Table 11.1: Instead of a zillion seldom-used API calls, the Renesas API set is short and sweet, making it very easy to get your arms around.

By now we are all very familiar with what it takes to establish and join a ZigBee PAN. After all, that's the whole idea behind this book, bringing an ocean of ZigBee and IEEE 802.15.4 information together in one small space, culling out the fat and providing you with ideas and stuff you can use to do your own ZigBee or IEEE 802.15.4 thing. Let's see if we can cut through the clutter and put a ZigBee PAN together using the Renesas API calls.

The first order of business in any ZigBee application is the initialization of the system. In the case of the M30280FA, a call to the M30280FA's SystemInit function not only sets up the M30280FA general-purpose I/O and M30280FA internal peripherals, the ZbReset function is also invoked as follows:

```
****************************************************************************
    result = ZbReset(0);
****************************************************************************
```

The value of 0 (zero) in the ZbReset function argument calls for a COLD reset. Here's what happens during the course of a COLD reset:

- An MLME-RESET.request(TRUE) is sent to the MAC layer

- The NWK NIB information entries are set to default values

- All control tables (neighbor tables, routing table, etc.) are cleared

If a WARM reset (function argument = 1) were requested, the only difference in execution is that the NIB information and control tables are left untouched. As indicated by the variable "result," the ZbReset function returns a result code. We are looking for a result code of 0x00, which says everything went down as it should during the reset operation. If you're wondering about the Renesas ZigBee stack and the RTOS components, they were initialized and started before the SystemInit function was called. The SystemInit function and everything else we will do with the application lies inside of a task, which is serviced by the RTOS.

We need to establish a PAN and to do that we first need to bring up a ZigBee Coordinator. However, first things first. You can't build a house without first buying some land. So, we call upon the ZbDiscoverNetwork API call, which is issued inside of a function called App_Discover, as follows:

```
****************************************************************************
    result = App_Discover(SCAN_CHANNELS, s_LocalPanId, &found);
****************************************************************************
```

which results in calling:

```
*****************************************************************************
    result = ZbDiscoverNetwork(ScanChannelsMap, // DWORD ScanChannels
            ZB_DEFAULT_SCAN_DURATION,          // BYTE ScanDuration
            Descs,                             // ZbPanDescriptor pDesc[]
            MAX_DESCRIPTOR_ENTRIES,            // BYTE DescSize
            &NetworkCount);// BYTE *pCount (number of entries filled by function)
*****************************************************************************
```

We are looking for existing networks here. As you know, there are none. (At least there should be none, as I don't have any running in the lab.) You also have figured out that I have limited the SCAN_CHANNELS value to channel 11 (0x00000800). The App_Discover function looks for a particular PanId. In our case the PanId we're searching for is the PanId we want to use for our PAN, which is 0x1ACE. If PanId 0x1ACE is found, the channel it was found on will be placed in the "found" variable. Since I'm rather sure we won't find a network at all, there will be no network count and no network descriptor entries this time around. The ZbDiscoverNetwork function will scan channel 11 for approximately 500 ms as the ZB_DEFAULT_SCAN_DURATION value is set for 0x05.

The ZbDiscoverNetwork API function will internally place a call to the MAC layer primitive MLME-SCAN.request and wait for a MLME-SCAN.confirm primitive. Any error returned by MLME-SCAN.confirm will be returned to the application. The Beacon Request shown in Sniffer Capture 11.1 was generated by the MAC scan as expected.

Sniffer Capture 11.1

```
*****************************************************************************
Frame 1 (Length = 10 bytes)
        Time Stamp: 15:22:43.000
        Frame Length: 10 bytes
        Capture Length: 10 bytes
        Link Quality Indication: 216
IEEE 802.15.4
        Frame Control: 0x0803
                .... .... .... .011 = Frame Type: Command (0x0003)
                .... .... .... 0... = Security Enabled: Disabled
                .... .... ...0 .... = Frame Pending: No more data
                .... .... ..0. .... = Acknowledgment Request: Acknowledgment
not required
                .... .... .0.. .... = Intra PAN: Not within the PAN
                .... ..00 0... .... = Reserved
                .... 10.. .... .... = Destination Addressing Mode: Address
field contains a 16-bit short address (0x0002)
                ..00 .... .... .... = Reserved
                00.. .... .... .... = Source Addressing Mode: PAN identifier
and address field are not present (0x0000)
        Sequence Number: 167
        Destination PAN Identifier: 0xffff
        Destination Address: 0xffff
        MAC Payload
```

```
          Command Frame Identifier = Beacon Request: (0x07)
     Frame Check Sequence: Correct

0000:   03 08 a7 ff ff ff ff 07 .. ..                    ..'.......
```
Sniffer Capture 11.1: No rocket science here that you don't already understand.

No networks were found. Thus, the next step is to start our ZigBee PAN on channel 11 and call it 0x1ACE. Here's the call:

```
   result = App_Start(DESIRED_CHANNEL,s_LocalPanId);
```

Which results in calling:

```
   result = ZbFormNetwork( ScanChannels, // DWORD ScanChannels
            ZB_DEFAULT_SCAN_DURATION,     // BYTE ScanDuratio
            ZB_DEFAULT_BEACON_ORDER,      // BYTE BeaconOrder (15=non-Beacon)
            ZB_DEFAULTSUPERFRM_ORDER,     // BYTE SuperFrameOrder(15=non-Beacon)
            PANId,                        // ZbPANId PANId
            TRUE);                        // BOOL ZbBatLifeExt
```

The ZbFormNetwork API function creates a PAN specified in PANId and the node that executes this API function assumes the ZigBee Coordinator position. A channel scan is performed by ZbFormNetwork to determine the best channel of the bunch to play in and to see if another PAN already exists on the channel. Thus, a second Beacon Request is issued and captured in Sniffer Capture 11.2.

Sniffer Capture 11.2

```
Frame 2 (Length = 10 bytes)
      Time Stamp: 15:22:44.019
      Frame Length: 10 bytes
      Capture Length: 10 bytes
      Link Quality Indication: 216
IEEE 802.15.4
      Frame Control: 0x0803
            .... .... .... .011  = Frame Type: Command (0x0003)
            .... .... .... 0...  = Security Enabled: Disabled
            .... .... ...0 ....  = Frame Pending: No more data
            .... .... ..0. ....  = Acknowledgment Request: Acknowledgment
not required
            .... .... .0.. ....  = Intra PAN: Not within the PAN
            .... ..00 0... ....  = Reserved
            .... 10.. .... ....  = Destination Addressing Mode: Address
field contains a 16-bit short address (0x0002)
            ..00 .... .... ....  = Reserved
            00.. .... .... ....  = Source Addressing Mode: PAN identifier
and address field are not present (0x0000)
```

```
        Sequence Number: 168
        Destination PAN Identifier: 0xffff
        Destination Address: 0xffff
        MAC Payload
                Command Frame Identifier = Beacon Request: (0x07)
        Frame Check Sequence: Correct


0000:    03 08 a8 ff ff ff ff 07 .. ..                        ..(.......
```
★★

Sniffer Capture 11.2: Don't fall asleep. The sequence numbers could be important if you're chasing through a sniff looking for a problem.

The newly anointed ZigBee Coordinator also broadcasts a Coordinator Realignment frame. That's sorta like a lion squirting on all of the trees in his territory. Frank Zappa, used to say "Don't eat the yellow snow." The yellow snow was captured by Daintree Networks SNA in Sniffer Capture 11.3.

Sniffer Capture 11.3

★★
```
Frame 3 (Length = 27 bytes)
        Time Stamp: 15:22:44.530
        Frame Length: 27 bytes
        Capture Length: 27 bytes
        Link Quality Indication: 216
IEEE 802.15.4
        Frame Control: 0xc803
                .... .... .... .011   = Frame Type: Command (0x0003)
                .... .... .... 0...   = Security Enabled: Disabled
                .... .... ...0 ....   = Frame Pending: No more data
                .... .... ..0. ....   = Acknowledgment Request: Acknowledgment
not required
                .... .... .0.. ....   = Intra PAN: Not within the PAN
                .... ..00 0... ....   = Reserved
                .... 10.. .... ....   = Destination Addressing Mode: Address
field contains a 16-bit short address (0x0002)
                ..00 .... .... ....   = Reserved
                11.. .... .... ....   = Source Addressing Mode: Address field
contains a 64-bit extended address (0x0003)
        Sequence Number: 169
        Destination PAN Identifier: 0xffff
        Destination Address: 0xffff
        Source PAN Identifier: 0x1ace
        Source Address: 0x11223344240000f6
        MAC Payload
                Command Frame Identifier = Coordinator Realignment: (0x08)
                PAN Identifier: 0x1ace
                Coordinator Short Address: 0x0000
                Logical Channel: 11
                Short Address: 0xffff
        Frame Check Sequence: Correct
```

```
0000:    03 c8 a9 ff ff ff ff ce 1a f6 00 00 24 44 33 22    .H).....N.v..$D3"
0010:    11 08 ce 1a 00 00 0b ff ff .. ..                   ..N.......
```

Sniffer Capture 11.3: This is totally unnecessary. However, it is a cool thing to do when a Sniffer is involved. Usually, this is sent to assist in rounding up orphan ZigBee nodes.

If a wanna-be ZigBee node can stand the smell and can tolerate yellow snow, the new ZigBee Coordinator is accepting guests as the following Renesas API function is initiated next:

```
    result = ZbPermitJoining(0xFF);        // (0xFF is always permitting)
```

You know the story. Right now, the new ZigBee PAN Coordinator is basking in the sun on his/her channel, holding court to wanna-be ZigBee nodes. If a wanna-be ZigBee node wants to come to the court of the crimson king, it must have initialized in the same manner as the ZigBee PAN Coordinator. I've got a Renesas M30280FA-based development board loaded with the Renesas ZigBee stack and the Renesas RTOS. Let's fire it up, turn it into a ZigBee End Device, and see if we can't wander into the king's court.

If we want to drop in on the king or queen, we must first find the castle. In this case, the castle is the PAN we know as 0x1ACE. I've already narrowed the channel search to make sure I catch everything flying about with the Daintree Networks SNA. The song remains the same as the wanna-be ZigBee End Device must scan channel 11 looking for an active ZigBee PAN. We can use the same Renesas API function we used as a ZigBee Coordinator to discover a network as a ZigBee End Device:

```
    result = App_Discover(SCAN_CHANNELS, s_LocalPanId, &found);
```

which results in calling:

```
    result = ZbDiscoverNetwork(ScanChannelsMap, // DWORD ScanChannels
            ZB_DEFAULT_SCAN_DURATION,           // BYTE ScanDuration
            Descs,                              // ZbPanDescriptor pDesc[]
            MAX_DESCRIPTOR_ENTRIES,             // BYTE DescSize
            &NetworkCount); // BYTE *pCount (number of entries filled by function)
```

The Beacon Request I captured in Sniffer Capture 11.4 was generated by the wanna-be Zig-Bee End Device as a result of the ZbDiscoverNetwork API call.

Sniffer Capture 11.4

```
Frame 4 (Length = 10 bytes)
        Time Stamp: 15:22:48.495
        Frame Length: 10 bytes
        Capture Length: 10 bytes
        Link Quality Indication: 224
IEEE 802.15.4
```

```
        Frame Control: 0x0803
                    .... .... .... .011  = Frame Type: Command (0x0003)
                    .... .... .... 0...  = Security Enabled: Disabled
                    .... .... ...0 ....  = Frame Pending: No more data
                    .... .... ..0. ....  = Acknowledgment Request: Acknowledgment
not required
                    .... .... .0.. ....  = Intra PAN: Not within the PAN
                    .... ..00 0... ....  = Reserved
                    .... 10.. .... ....  = Destination Addressing Mode: Address
field contains a 16-bit short address (0x0002)
                    ..00 .... .... ....  = Reserved
                    00.. .... .... ....  = Source Addressing Mode: PAN identifier
and address field are not present (0x0000)
        Sequence Number: 106
        Destination PAN Identifier: 0xffff
        Destination Address: 0xffff
        MAC Payload
                Command Frame Identifier = Beacon Request: (0x07)
        Frame Check Sequence: Correct

0000:   03 08 6a ff ff ff ff 07 .. ..                    ..j.......
```

****

Sniffer Capture 11.4: London Calling…(A must-have album released by The Clash in the UK in 1979).

The ZigBee Coordinator is accepting nodes into the PAN and responds with the informational Beacon caught in Sniffer Capture 11.5.

Sniffer Capture 11.5

****

```
Frame 5 (Length = 16 bytes)
        Time Stamp: 15:22:48.497
        Frame Length: 16 bytes
        Capture Length: 16 bytes
        Link Quality Indication: 216
IEEE 802.15.4
        Frame Control: 0x8000
                    .... .... .... .000  = Frame Type: Beacon (0x0000)
                    .... .... .... 0...  = Security Enabled: Disabled
                    .... .... ...0 ....  = Frame Pending: No more data
                    .... .... ..0. ....  = Acknowledgment Request: Acknowledgment
not required
                    .... .... .0.. ....  = Intra PAN: Not within the PAN
                    .... ..00 0... ....  = Reserved
                    .... 00.. .... ....  = Destination Addressing Mode: PAN
identifier and address field are not present (0x0000)
                    ..00 .... .... ....  = Reserved
                    10.. .... .... ....  = Source Addressing Mode: Address field
contains a 16-bit short address (0x0002)
        Sequence Number: 180
        Source PAN Identifier: 0x1ace
```

```
        Source Address: 0x0000
        MAC Payload
              Superframe Specification: 0xdfff
                           .... .... .... 1111  = Beacon Order (0x000f)
                           .... .... 1111 ....  = Superframe Order (0x000f)
                           .... 1111 .... ....  = Final CAP Slot (0x000f)
                           ...1 .... .... ....  = Battery Life Extension: Enabled
                           ..0. .... .... ....  = Reserved
                           .1.. .... .... ....  = PAN Coordinator: Transmitter is
a PAN Coordinator
                           1... .... .... ....  = Association Permit: Coordinator
accepting Association Requests
              GTS Specification: 0x80
                           .... .000  = GTS Descriptor Count (0x00)
                           .000 0...  = Reserved
                           1... ....  = GTS Permit: Coordinator accepting GTS
Requests
              Pending Address Specification: 0x00
                           .... .000  = Number of short Addresses pending: 0
                           .... 0...  = Reserved
                           .000 ....  = Number of extended Addresses pending: 0
                           0... ....  = Reserved
              Beacon Payload
                    Protocol ID: ZigBee NWK (0x00)
        Frame Check Sequence: Correct
NWK Layer Information: 0x8411
        .... .... .... 0001  = Stack Profile (0x1)
        .... .... 0001 ....  = nwkcProtocolVersion (0x1)
        .... ..00 .... ....  = Reserved (0x0)
        .... .1.. .... ....  = Router Capacity: True
        .000 0... .... ....  = Device Depth (0x0)
        1... .... .... ....  = End Device Capacity: True

0000:   00 80 b4 ce 1a 00 00 ff df 80 00 00 11 84 .. ..      ..4N...._.......
```
****
Sniffer Capture 11.5: The most important portion of this message to a needy ZigBee End Device is the very last line of this sniff.

This time around, a ZigBee PAN is operational on channel 11, the informational Beacon was transmitted by the ZigBee Coordinator, and the "found" variable is filled with the value of the channel on which the discovered ZigBee PAN is operating. Normally, the s_LocalPanId value would be the PAN ID of the ZigBee PAN the ZigBee End Device has chosen to join following a discovery. The concept is important here and to keep it simple (the way I like it), we'll use the well-known PanId value of 0x1ACE (which just happens to be the same as the discovered value) and call the ZbJoin API function like this:

****
```
result = App_Join(found,s_LocalPanId );
```
****

Which results in calling:

```
**************************************************************************
        result = ZbJoin(s_LocalPanId,        // ZbPANId PANId
        FALSE,                                // BOOL JoinAsRn
        FALSE,                                // BOOL RejoinNet
        ScanChannels,                         // DWORD ScanChannels
        ZB_DEFAULT_SCAN_DURATION,             // BYTE ScanDuration
        0,                                    // BYTE PowerSrc(0: mains, 1:alt)
        rx_on_idle,                           // BOOL RxOnWhenIdle
        0);                                   // BOOL MacSec
**************************************************************************
```

The parameters that are stuffed into the ZbJoin API call are rather easy to comprehend. We're not joining the ZigBee PAN as a Router, thus that parameter is FALSE and we're not rejoining the PAN, which requires a FALSE value as well. The ScanChannels and scan duration values have not changed from the previous values you were given and security is a no-go. Since we formed a nonBeacon network, we can leave the receiver on to accept messages whenever they are sent our way. It helps to let the ZigBee PAN Coordinator know our intentions and we do so in Sniffer Capture 11.6.

Sniffer Capture 11.6

```
**************************************************************************
Frame 6 (Length = 21 bytes)
        Time Stamp: 15:22:49.008
        Frame Length: 21 bytes
        Capture Length: 21 bytes
        Link Quality Indication: 224
IEEE 802.15.4
        Frame Control: 0xc823
                .... .... .... .011 = Frame Type: Command (0x0003)
                .... .... .... 0... = Security Enabled: Disabled
                .... .... ...0 .... = Frame Pending: No more data
                .... .... ..1. .... = Acknowledgment Request: Acknowledgment
required
                .... .... .0.. .... = Intra PAN: Not within the PAN
                .... ..00 0... .... = Reserved
                .... 10.. .... .... = Destination Addressing Mode: Address
field contains a 16-bit short address (0x0002)
                ..00 .... .... .... = Reserved
                11.. .... .... .... = Source Addressing Mode: Address field
contains a 64-bit extended address (0x0003)
        Sequence Number: 107
        Destination PAN Identifier: 0x1ace
        Destination Address: 0x0000
        Source PAN Identifier: 0xffff
        Source Address: 0x11223344240000f5
        MAC Payload
                Command Frame Identifier = Association Request: (0x01)
                Capability Information: 0x88
                        .... ...0 = Alternate PAN Coordinator: Not capable of
becoming PAN Coordinator
                        .... ..0. = Device Type: RFD
```

```
                          .... .0..  = Power Source: Not receiving power from
alternating current mains
                          .... 1...  = Receiver on when idle: Enables receiver
when idle
                          ..00 ....  = Reserved
                          .0.. ....  = Security Capability: Not capable of using
security suite
                          1... ....  = Allocate Address: Coordinator should
allocate short address
        Frame Check Sequence: Correct


0000:   23 c8 6b ce 1a 00 00 ff ff f5 00 00 24 44 33 22     #HkN.....u..$D3"
0010:   11 01 88 .. ..                                      .....
```
****

Sniffer Capture 11.6: The most important part of this message as far as the ZigBee End Device is concerned is the Allocate Address request. Once an address is assigned to the wanna-be ZigBee End Device, the party in the court of the crimson king is on.

If you actually read the sniff carefully, you know that an acknowledgment frame is coming and here it is:

****
```
Frame 7 (Length = 5 bytes)
        Time Stamp: 15:22:49.009
        Frame Length: 5 bytes
        Capture Length: 5 bytes
        Link Quality Indication: 216
IEEE 802.15.4
        Frame Control: 0x0002
                  .... .... .... .010  = Frame Type: Acknowledgment (0x0002)
                  .... .... .... 0...  = Security Enabled: Disabled
                  .... .... ...0 ....  = Frame Pending: No more data
                  .... .... ..0. ....  = Acknowledgment Request: Acknowledgment
not required
                  .... .... .0.. ....  = Intra PAN: Not within the PAN
                  .... ..00 0... ....  = Reserved
                  .... 00.. .... ....  = Destination Addressing Mode: PAN
identifier and address field are not present (0x0000)
                  ..00 .... .... ....  = Reserved
                  00.. .... .... ....  = Source Addressing Mode: PAN identifier
and address field are not present (0x0000)
        Sequence Number: 107
        Frame Check Sequence: Correct


0000:   02 00 6b .. ..                                      ..k..
```
****

The ZigBee End Device is still waiting for a confirmation of the association and deals out a Data Request frame like this:

****
```
Frame 8 (Length = 18 bytes)
```

```
        Time Stamp: 15:22:49.502
        Frame Length: 18 bytes
        Capture Length: 18 bytes
        Link Quality Indication: 224
IEEE 802.15.4
        Frame Control: 0xc863
                .... .... .... .011  = Frame Type: Command (0x0003)
                .... .... .... 0...  = Security Enabled: Disabled
                .... .... ...0 ....  = Frame Pending: No more data
                .... .... ..1. ....  = Acknowledgment Request: Acknowledgment
required
                .... .... .1.. ....  = Intra PAN: Within the PAN
                .... ..00 0... ....  = Reserved
                .... 10.. .... ....  = Destination Addressing Mode: Address
field contains a 16-bit short address (0x0002)
                ..00 .... .... ....  = Reserved
                11.. .... .... ....  = Source Addressing Mode: Address field
contains a 64-bit extended address (0x0003)
        Sequence Number: 108
        Destination PAN Identifier: 0x1ace
        Destination Address: 0x0000
        Source Address: 0x11223344240000f5
        MAC Payload
                Command Frame Identifier = Data Request: (0x04)
        Frame Check Sequence: Correct

0000:   63 c8 6c ce 1a 00 00 f5 00 00 24 44 33 22 11 04     cHlN...u..$D3"..
0010:   .. ..                                               ..
```
****

One more acknowledgment:

****
```
Frame 9 (Length = 5 bytes)
        Time Stamp: 15:22:49.503
        Frame Length: 5 bytes
        Capture Length: 5 bytes
        Link Quality Indication: 216
IEEE 802.15.4
        Frame Control: 0x0012
                .... .... .... .010  = Frame Type: Acknowledgment (0x0002)
                .... .... .... 0...  = Security Enabled: Disabled
                .... .... ...1 ....  = Frame Pending: More data
                .... .... ..0. ....  = Acknowledgment Request: Acknowledgment
not required
                .... .... .0.. ....  = Intra PAN: Not within the PAN
                .... ..00 0... ....  = Reserved
                .... 00.. .... ....  = Destination Addressing Mode: PAN
identifier and address field are not present (0x0000)
                ..00 .... .... ....  = Reserved
                00.. .... .... ....  = Source Addressing Mode: PAN identifier
and address field are not present (0x0000)
        Sequence Number: 108
```

```
        Frame Check Sequence: Correct

0000:   12 00 6c .. ..                              ..1..
**********************************************************************
```

And, the wanna-be ZigBee End Device, now known as 0x1558, is now a courtier of the king:
```
**********************************************************************
Frame 10 (Length = 27 bytes)
        Time Stamp: 15:22:49.504
        Frame Length: 27 bytes
        Capture Length: 27 bytes
        Link Quality Indication: 216
IEEE 802.15.4
        Frame Control: 0xcc63
                .... .... .... .011 = Frame Type: Command (0x0003)
                .... .... .... 0... = Security Enabled: Disabled
                .... .... ...0 .... = Frame Pending: No more data
                .... .... ..1. .... = Acknowledgment Request: Acknowledgment
required
                .... .... .1.. .... = Intra PAN: Within the PAN
                .... ..00 0... .... = Reserved
                .... 11.. .... .... = Destination Addressing Mode: Address
field contains a 64-bit extended address (0x0003)
                ..00 .... .... .... = Reserved
                11.. .... .... .... = Source Addressing Mode: Address field
contains a 64-bit extended address (0x0003)
        Sequence Number: 170
        Destination PAN Identifier: 0x1ace
        Destination Address: 0x11223344240000f5
        Source Address: 0x11223344240000f6
        MAC Payload
                Command Frame Identifier = Association Response: (0x02)
                Short Address: 0x1558
                Association Status: Association Successful (0x00)
        Frame Check Sequence: Correct

0000:   63 cc aa ce 1a f5 00 00 24 44 33 22 11 f6 00 00    cL*N.u..$D3".v..
0010:   24 44 33 22 11 02 58 15 00 .. ..                   $D3"..X....
**********************************************************************
```

We need to acknowledge the acknowledgment as there is more:
```
**********************************************************************
Frame 11 (Length = 5 bytes)
        Time Stamp: 15:22:49.505
        Frame Length: 5 bytes
        Capture Length: 5 bytes
        Link Quality Indication: 224
IEEE 802.15.4
        Frame Control: 0x0002
                .... .... .... .010 = Frame Type: Acknowledgment (0x0002)
                .... .... .... 0... = Security Enabled: Disabled
                .... .... ...0 .... = Frame Pending: No more data
```

```
            .... .... ..0. ....  = Acknowledgment Request: Acknowledgment
not required
            .... .... .0.. ....  = Intra PAN: Not within the PAN
            .... ..00 0... ....  = Reserved
            .... 00.. .... ....  = Destination Addressing Mode: PAN
identifier and address field are not present (0x0000)
            ..00 .... .... ....  = Reserved
            00.. .... .... ....  = Source Addressing Mode: PAN identifier
and address field are not present (0x0000)
      Sequence Number: 170
      Frame Check Sequence: Correct


0000:    02 00 aa .. ..                                    ..*..
****************************************************************************
```

Let's throw some data around to see how the ZbDataSnd API function works. We'll send three bytes of information to the ZigBee Pan Coordinator in an announcement message intended to let provide some information for the ZigBee Pan Coordinator to store away in a database for later use. Here's the M30280FA code:

```
****************************************************************************
    tx_buffer[0] = 1;              // 1 means a new device has joined
    tx_buffer[1] = (unsigned char)(my_nwk_address >> 8);
    tx_buffer[2] = (unsigned char)my_nwk_address;
    result = ZbDataSnd(55,         // ZbNsduHandle handle,
    0x0000,                        // pZbShortAddr pDstAddr,
    tx_buffer,                     // BYTE FAR *pData,
    3,                             // int len,
    15,                            // BYTE Radius,
    1,                             // BYTE RouteDiscovery (0: Along Tree,
1:Discover if needed, 2: Force Discovery)
    FALSE,                         // BOOL Security,
    0);                            // BYTE option=0x00 );
****************************************************************************
```

The ZbNsduHandle value is an arbitrary value intended to be used when multiple calls come in from differing tasks. This version of the Renesas API doesn't support the multiple-call functionality and you won't see the ZbNsduHandle value in the sniff. If you're wondering where my_nwk_address value came from, it was retrieved immediately following the successful association with this line of code:

```
****************************************************************************
ZbShortAddr_Get(&my_nwk_address); // Retrieve short address from stack
****************************************************************************
```

The ZigBee frame is transmitted by the new ZigBee End Device and, as you can see in Sniffer Capture 11.7, I was ready for it.

Sniffer Capture 11.7

```
****************************************************************************
Frame 12 (Length = 22 bytes)
      Time Stamp: 15:22:49.523
```

```
                Frame Length: 22 bytes
                Capture Length: 22 bytes
                Link Quality Indication: 236
IEEE 802.15.4
        Frame Control: 0x8861
                .... .... .... .001  = Frame Type: Data (0x0001)
                .... .... .... 0...  = Security Enabled: Disabled
                .... .... ...0 ....  = Frame Pending: No more data
                .... .... ..1. ....  = Acknowledgment Request: Acknowledgment
required
                .... .... .1.. ....  = Intra PAN: Within the PAN
                .... ..00 0... ....  = Reserved
                .... 10.. .... ....  = Destination Addressing Mode: Address
field contains a 16-bit short address (0x0002)
                ..00 .... .... ....  = Reserved
                10.. .... .... ....  = Source Addressing Mode: Address field
contains a 16-bit short address (0x0002)
        Sequence Number: 109
        Destination PAN Identifier: 0x1ace
        Destination Address: 0x0000
        Source Address: 0x1558
        Frame Check Sequence: Correct
ZigBee NWK
        Frame Control: 0x0044
                .... .... .... ..00  = Frame Type: NWK Data (0x00)
                .... .... ..00 01..  = Protocol Version (0x01)
                .... .... 01.. ....  = Discover Route: Enable route discovery
(0x01)
                .... ...0 .... ....  = Reserved
                .... ..0. .... ....  = Security: Disabled
                0000 00.. .... ....  = Reserved
        Destination Address: 0x0000
        Source Address: 0x1558
        Radius = 15
        Sequence Number = 133
NWK Payload: 01:15:58

0000:   61 88 6d ce 1a 00 00 58 15 44 00 00 00 58 15 0f    a.mN...X.D...X..
0010:   85 01 15 58 .. ..                                  ...X..
```

****

Sniffer Capture 11.7: I think you see how you can put your data into a ZigBee frame that is generated from a simple little application talking to the NWK layer. Our three bytes of data are decoded at the end of this sniff.

In this chapter, we crossed over ever so slightly into the land of ZigBee. The Renesas API set made that a relatively easy trip. You are now NWK qualified. You and I have proven that we can indeed generate a message from above the IEEE 802.15.4 clouds.

About Renesas

When I see the name Renesas, I immediately think high-performance microcontrollers. I was pleasantly surprised with the quality and quantity of IEEE 802.15.4 and ZigBee stuff Renesas is offering. Tim Dry of Renesas was also one of the very first contributors to the content of this book. Thank you, Tim. In fact, Renesas's early contributions in the writing cycle allowed me to see the worth of the Renesas IEEE 802.15.4-compliant/ZigBee-ready solutions, as I used the Renesas RF Sniffer board to capture 99% of the 2.4-GHz sniffs in the pages of this book.

My teen years (yes, I'm old) were inundated with the music of Motown. The Temptations I grew up with were made up of Melvin Franklin, Eddie Kendricks, Otis Williams, Paul Williams, and the baddest man on the planet at that time, David Ruffin. David mesmerized the world with the Temps 1965 hit, "My Girl."

If you can't answer this musical question, give this book to your sister. What female vocal group became the very first group to have four number-one singles in a row? Hint: They also made a record or two with the Temps.

Freescale

I love Freescale Semiconductor. All of the Freescale Semiconductor development kits I've had the pleasure of evaluating lately have come in very pretty "drop-me, spill-crap-on-me" cases. One such example is the MC13213-based IEEE 802.15.4/ZigBee development platform I shot (shot in this sense is photographed for those of you that hunt) for you in Photo 12.1.

Photo 12.1: MC13213-based IEEE 802.15.4/ZigBee development platform. A peek through the looking glass. Everything in this shot is under control of the components inside the little outlined box directly under the BDM connector.

The big black square within the silkscreened box to the top right of the LCD is a Freescale Semiconductor MC13213. The MC13213 RF transceiver is a full-blown IEEE 802.15.4-compliant radio operating in the 2.4-GHz ISM frequency band. The transceiver includes a low-noise amplifier, 1-mW nominal output power, voltage controlled oscillator (VCO), integrated transmit/receive switch, on-board power supply regulation, and full spread-spectrum encoding and decoding. Nothing unlike what we've seen up to this point as far as IEEE 802.15.4-compliant transceivers are concerned. However, the MC13213 is not only an IEEE 802.15.4/ZigBee radio. There's a Freescale Semiconductor microcontroller in there too. The microcontroller portion of the MC13213 is based on the Freescale Semiconductor HCS08 family of microcontrollers. The MC1321x family is Freescale Semiconductor's second-generation ZigBee platform. The MC13213 contains 60K of Flash and 4 KB of RAM and is intended for use with the Freescale fully compliant 802.15.4 MAC. Some say (including Freescale Semiconductor) that the concept behind the official IEEE 802.15.4 PHY we've been running through is a product of Freescale Semiconductor engineering. Running up the ZigBee stack, the next layer is…say it…the MAC, and guess what? Freescale Semiconductor and Royal Philips (I love kingsy and queensy stuff the British do—the food ain't bad either) provided the basis for the official IEEE 802.15.4 MAC as we know it.

It's rather obvious that the Freescale Semiconductor development platform in Photo 12.1 is assuming the role of PAN Coordinator. No single-node PAN is an island. So, the electronics-in-a-box in Photo 12.2 holds up the Router and ZigBee End Device sector of the PAN.

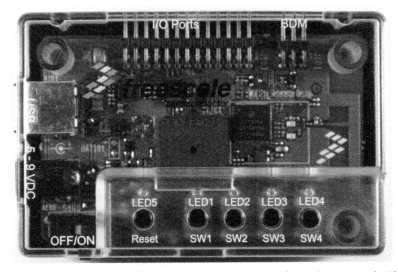

Photo 12.2: There's not much difference in this unit versus the unit you see in Photo 12.1. As you can see, this unit does not include an RS-232 interface.

Regardless of the functionality of the IEEE 802.15.4/ZigBee platforms you see in Photos 12.1 and 12.2, the show is under the control of MC13213, which you see enclosed in a silkscreen box in Photos 12.1 and 12.2 and up close in Photo 12.3.

One other goody I found handy while working with the MC13213-based development platforms was the PEmicro USB Multilink Interface. During my initial experimentation with the MC13213 development platforms, I managed to "lose" some of the code while attempting to upload an application from my laptop to the MC13213-based PAN Coordinator. After some number of futile attempts to regain control without the PEmicro USB Multilink Interface pod, I finally broke down and installed the PEmicro BDM unit and successfully brought myself back to a place of comfort within the Freescale Semiconductor IEEE 802.15.4 framework. My hero posed for the shot in Photo 12.4.

Freescale Semiconductor's hardware-development platforms are always beautiful to behold. However, there's code that accompanies the designer hardware. So, let's go see how the folks at Freescale Semiconductor do IEEE 802.15.4 networking.

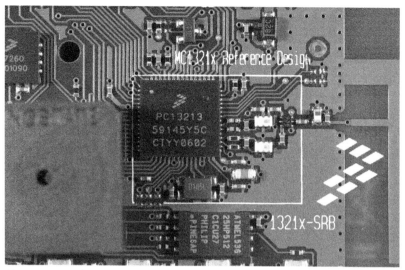

Photo 12.3: I love the B2 Bomber logo.

Photo 12.4: I've been a fan of this little guy's big brother for quite some time. I've done many a magazine column using the PEmicro Cyclone Pro.

Instead of tracking the code and matching it up to a Daintree Networks SNA capture, let's do something different this time around. Let's take a complete Daintree Networks SNA capture and match up the frames to the SMAC (Simple MAC) primitive calls that spawned them. The PAN Coordinator scanned the entire 2.4-GHz range of channels. I managed to finally find the PAN it established on channel 26 and obtained a full capture of the events that transpired. In that I did not see any Beacon Requests generated while the PAN Coordinator was searching the 2.4-GHz channel set for a suitable channel to start its PAN, I must assume that a passive ED (Energy Detect) scan was used instead of an active scan. It found the operational channel by searching for an End Device association frame sequence. The End Device used an active scan technique to find the PAN Coordinator. Sniffer Capture 12.1 is a textual depiction of the initial End Device Beacon Request frame I captured on channel 24:

Sniffer Capture 12.1

```
************************************************************************
Frame 1 (Length = 10 bytes)
        Time Stamp: 12:21:17.000
        Frame Length: 10 bytes
        Capture Length: 10 bytes
        Link Quality Indication: 152
IEEE 802.15.4
        Frame Control: 0x0803
                .... .... .... .011  = Frame Type: Command (0x0003)
                .... .... .... 0...  = Security Enabled: Disabled
                .... .... ...0 ....  = Frame Pending: No more data
                .... .... ..0. ....  = Acknowledgment Request: Acknowledgment
not required
                .... .... .0.. ....  = Intra PAN: Not within the PAN
                .... ..00 0... ....  = Reserved
                .... 10.. .... ....  = Destination Addressing Mode: Address
field contains a 16-bit short address (0x0002)
                ..00 .... .... ....  = Reserved
                00.. .... .... ....  = Source Addressing Mode: PAN identifier
and address field are not present (0x0000)
        Sequence Number: 237
        Destination PAN Identifier: 0xffff
        Destination Address: 0xffff
        MAC Payload
                Command Frame Identifier = Beacon Request: (0x07)
        Frame Check Sequence: Correct

0000:   03 08 ed ff ff ff ff 07 .. ..              ..m.......
************************************************************************
```

Sniffer Capture 12.1: No bit has been left unturned. Without Daintree Networks SNA, this would be an "unknown" as the radio isn't going to send you an email telling you which channel it has selected.

As my doctor would say, there's nothing remarkable about the Beacon Request that is Frame 1 of this exploratory capture. However, the response frame from the PAN Coordinator gives us some clues as to what the PAN is capable of. Examine the contents of Sniffer Capture 12.2:

Sniffer Capture 12.2

```
************************************************************************
Frame 2 (Length = 13 bytes)
        Time Stamp: 12:21:17.004
        Frame Length: 13 bytes
        Capture Length: 13 bytes
        Link Quality Indication: 136
```

```
IEEE 802.15.4
      Frame Control: 0x8000
                  .... .... .... .000  = Frame Type: Beacon (0x0000)
                  .... .... .... 0...  = Security Enabled: Disabled
                  .... .... ...0 ....  = Frame Pending: No more data
                  .... .... ..0. ....  = Acknowledgment Request: Acknowledgment
not required
                  .... .... .0.. ....  = Intra PAN: Not within the PAN
                  .... ..00 0... ....  = Reserved
                  .... 00.. .... ....  = Destination Addressing Mode: PAN
identifier and address field are not present (0x0000)
                  ..00 .... .... ....  = Reserved
                  10.. .... .... ....  = Source Addressing Mode: Address field
contains a 16-bit short address (0x0002)
      Sequence Number: 84
      Source PAN Identifier: 0xaaaa
      Source Address: 0xcafe
      MAC Payload
            Superframe Specification: 0xcfff
                  .... .... .... 1111  = Beacon Order (0x000f)
                  .... .... 1111 ....  = Superframe Order (0x000f)
                  .... 1111 .... ....  = Final CAP Slot (0x000f)
                  ...0 .... .... ....  = Battery Life Extension: Disabled
                  ..0. .... .... ....  = Reserved
                  .1.. .... .... ....  = PAN Coordinator: Transmitter is
a PAN Coordinator
                  1... .... .... ....  = Association Permit: Coordinator
accepting Association Requests
            GTS Specification: 0x00
                  .... .000  = GTS Descriptor Count (0x00)
                  .000 0...  = Reserved
                  0... ....  = GTS Permit: Coordinator not accepting GTS
Requests
            Pending Address Specification: 0x00
                  .... .000  = Number of short Addresses pending: 0
                  .... 0...  = Reserved
                  .000 ....  = Number of extended Addresses pending: 0
                  0... ....  = Reserved
      Frame Check Sequence: Correct

0000:   00 80 54 aa aa fe ca ff cf 00 00 .. ..          ..T**~J.O....
```
****

Sniffer Capture 12.2: Don't expect an email with all of this information either. If the End Device wants to join this PAN, it has the keys to the door.

Just a cursory look at the capture data in Frame 2 tells us that the PAN identifier is 0x0AAAA and we may address the PAN Coordinator as 0x0CAFE. The Beacon Order and Superframe Order bytes indicate a nonBeacon network, which is accepting the possible association of applicants.

Let's stop here and catch the MC13213 code execution up with the Daintree Networks SNA sniffs. Code Snippet 12.1 is the code behind the PAN identifier and the PAN Coordinator's Source Address:

Code Snippet 12.1

```
*********************************************************************
/* We want the coordinators short address to be 0xCAFE. */
const uint8_t shortAddress[2] = { 0xFE, 0xCA };

/* PAN ID is 0xAAAA */
const uint8_t panId[2] = { 0xAA, 0xAA };
*********************************************************************
```
Code Snippet 12.1: Nothing fancy here. Just simple C constant declarations.

The code behind the capture is based on a simple state machine concept. Before any execution of code statements that reside in the application's main loop, the initial state is coded to stateInit. The states are traversed as shown in Code Snippet 12.2:

Code Snippet 12.2

```
*********************************************************************
enum
{
  stateInit,
  stateScanEdStart,
  stateScanEdWaitConfirm,
  stateStartCoordinator,
  stateStartCoordinatorWaitConfirm,
  stateListen,
  stateTerminate
};
*********************************************************************
```
Code Snippet 12.2: I really get perturbed when readers tell me that C source code is hard to follow and understand. The code here not only tells you it is a "state," but it spells out the state type in plain English.

My Daintree Networks SNA Packet List window is empty at this point. The basic initialization tasks must first be performed and that is done while the application state is statiInit. As you can see in the Code Snippet 12.3, the 802.15.4 stack in initialized and the state machine is advanced:

Code Snippet 12.3

```
*********************************************************************
    switch(state)
    {
    case stateInit:
      /* Initialize the 802.15.4 stack */
      Init_802_15_4();

      /* Goto Energy Detection state. */
```

```
    state = stateScanEdStart;

    break;
```
****
Code Snippet 12.3: No rocket science here.

The Init_802_15_4 function resets state machines and initializes internal module variables. When the dust settles, the MAC and PHY layer services are available to the application. The main application loop will eventually return and walk through the Select statement tree to the case of the current state, which is stateScanEdStart as shown in Code Snippet 12.4:

Code Snippet 12.4
****
```
    case stateScanEdStart:
      /* Start the Energy Detection scan, and goto wait for confirm state. */
      ret = App_StartScan(gScanModeED_c);
      if(ret == errorNoError)
      {
        state = stateScanEdWaitConfirm;
      }
      break;
```
****
Code Snippet 12.4: This piece of code begins the process of finding some RF out there.

The App_StartScan functions passes the scanType argument to a message that will be sent to the MAC via the management SAP that interfaces the NWK and MAC layers. The application has taken upon itself to act as an NWK layer when necessary. In the App_StartScan code that follows, note that the SCAN_CHANNELS value is predefined to allow the MC13213 to scan all of the 2.4-GHz channel space. The ED scan checks for radiation levels on all of the scanned channels. The application will take the returned radiation data and choose the channel with the amount of radiated energy as the PAN Coordinator's spawn channel. I've been dancing around scan duration figures since they were not relevant, as I locked in on a single channel for scanning. The App_StartScan code in Code Snippet 12.5 provides a good look at the match behind the scan duration parameter:

Code Snippet 12.5
****
```
#define SCAN_CHANNELS           0x07FFF800

uint8_t App_StartScan(uint8_t scanType)
{
  mlmeMessage_t *pMsg;
  mlmeScanReq_t *pScanReq;

  /* Allocate a message for the MLME (We should check for NULL). */
  pMsg = MSG_AllocType(mlmeMessage_t);
  if(pMsg != NULL)
  {
```

```
    /* This is a MLME-START.req command */
    pMsg->msgType = gMlmeScanReq_c;
    /* Create the Start request message data. */
    pScanReq = &pMsg->msgData.scanReq;
    /* gScanModeED_c, gScanModeActive_c, gScanModePassive_c, or
gScanModeOrphan_c */
    pScanReq->scanType = scanType;
    /* ChannelsToScan & 0xFF - LSB, always 0x00 */
    pScanReq->scanChannels[0] = (uint8_t)((SCAN_CHANNELS)      & 0xFF);
    /* ChannelsToScan>>8 & 0xFF   */
    pScanReq->scanChannels[1] = (uint8_t)((SCAN_CHANNELS>>8)   & 0xFF);
    /* ChannelsToScan>>16 & 0xFF   */
    pScanReq->scanChannels[2] = (uint8_t)((SCAN_CHANNELS>>16)  & 0xFF);
    /* ChannelsToScan>>24 & 0xFF - MSB */
    pScanReq->scanChannels[3] = (uint8_t)((SCAN_CHANNELS>>24)  & 0xFF);
    /* Duration per channel 0-14 (dc). T[sec] = (16*960*((2^dc)+1))/1000000.
       A scan duration of 5 on 16 channels approximately takes 8 secs. */
    pScanReq->scanDuration = 5;

    /* Send the Scan request to the MLME. */
    if(MSG_Send(NWK_MLME, pMsg) == gSuccess_c)
    {
      return errorNoError;
    }
  }
}
```

****
Code Snippet 12.5: As you can see, scanning around the spectrum takes quite a bit of time in the 802.15.4 domain.

Along the way we have learned that primitives can have up to three states of existence, which are:

- Request
- Confirm
- Indication

We just kicked off an MLME-SCAN.request primitive and you can bet that a confirmation originating from the MLME is in the works. The primitive messages are double queued (queued in both directions) between the NWK and MLME layers to avoid bogging down the system. The SAP handler performs the queueing and does not process the primitive messages it is passing in any way.

Once the NWK message with a message type of gNwkScanCnf_c is received, the scan confirmation message can be processed. The code in Code Snippet 12.6 performs the channel scan and chooses a suitable channel on which to establish the PAN:

Code Snippet 12.6

```
*****************************************************************************
    case stateScanEdWaitConfirm:
      /* Stay in this state until the MLME Scan confirm message arrives,
         and has been processed. Then goto Start Coordinator state. */
      ret = App_WaitMsg(pMsgIn, gNwkScanCnf_c);
      if(ret == errorNoError)
      {
        /* Process the ED scan confirm. The logical
           channel is selected by this function. */
        App_HandleScanEdConfirm(pMsgIn);
        state = stateStartCoordinator;
      }
      break;
*****************************************************************************
```

Code Snippet 12.6: This is where Spock would say "Scanning, Captain."

The downside to using energy detection instead of active scanning is that the channel could be occupied by a sleeping PAN Coordinator. We could have also scanned at just the right time when everyone on the PAN was silent. That's why the length of the scan duration is important when using the energy detection scan method. If you persist on the channel for long enough, you will most likely hear someone if they are really there.

The channel select code in Code Snippet 12.7 is rather clever. Once a real message has been detected by the App_WaitMsg function, here's what transpires:

Code Snippet 12.7

```
*****************************************************************************
void App_HandleScanEdConfirm(nwkMessage_t *pMsg)
{
  uint8_t n, minEnergy;
  uint8_t *pEdList;

  /* Get a pointer to the energy detect results */
  pEdList = pMsg->msgData.scanCnf.resList.pEnergyDetectList;

  /* Set the minimum energy to a large value */
  minEnergy = 0xFF;

  /* Select default channel */
  logicalChannel = 11;

  /* Search for the channel with least energy */
  for(n=0; n<16; n++)
  {
    if(pEdList[n] < minEnergy)
    {
      minEnergy = pEdList[n];
      /* Channel numbering is 11 to 26 both inclusive */
      logicalChannel = n + 11;
```

```
    }
  }

  /* The list of detected energies must be freed. */
  MSG_Free(pEdList);
}
```

Code Snippet 12.7: This is like television channel surfing but instead of looking a really good program or a great signal, we're looking the channel with the least amount of signal.

The MLME-SCAN.confirm primitive returns a list of the scanned channels' energy levels ordered from channel 11 (pEdList[0]) through channel 26 (pEdList[15]). Each channel energy level in the list is compared and the channel with the least detected energy is selected by applying its list offset to the default (beginning) channel of 11. The quiet channel then becomes the logicalChannel. With the operational channel officially selected, the application can move into the next state, which is coded up in Code Snippet 12.8, stateStartCoordinator:

Code Snippet 12.8

```
    case stateStartCoordinator:
      /* Start up as a PAN Coordinator on the selected channel. */
      ret = App_StartCoordinator();
      if(ret == errorNoError)
      {
        /* If the Start request was sent successfully to
           the MLME, then goto Wait for confirm state. */
        state = stateStartCoordinatorWaitConfirm;
      }
      break;
```

Code Snippet 12.8: Let her rip!

Some examples of PAN initiation we have seen reverberated a Coordinator Realignment frame to announce the new PAN attributes and allow us a preview of them on a Daintree Networks SNA sniff. In this case, Frame 2 is a validation of the App_StartCoordinator code in Code Snippet 12.9:

Code Snippet 12.9

```
uint8_t App_StartCoordinator(void)
{
  /* Message for the MLME will be allocated and attached to this pointer */
  mlmeMessage_t *pMsg;

  /* Allocate a message for the MLME (We should check for NULL). */
  pMsg = MSG_AllocType(mlmeMessage_t);
  if(pMsg != NULL)
  {
```

```
/* Pointer which is used for easy access inside the allocated message */
mlmeStartReq_t *pStartReq;
/* Return value from MSG_send - used for avoiding compiler warnings */
uint8_t ret;
/* Boolean value that will be written to the MAC PIB */
uint8_t boolFlag;

/* Set-up MAC PIB attributes. Please note that Set, Get,
   and Reset messages are not freed by the MLME. */

/* We must always set the short address to something
   other than 0xFFFF before starting a PAN. */
pMsg->msgType = gMlmeSetReq_c;
pMsg->msgData.setReq.pibAttribute = gMacPibShortAddress_c;
pMsg->msgData.setReq.pibAttributeValue = (uint8_t *)shortAddress;
ret = MSG_Send(NWK_MLME, pMsg);

/* We must set the Association Permit flag to TRUE
   in order to allow devices to associate to us. */
pMsg->msgType = gMlmeSetReq_c;
pMsg->msgData.setReq.pibAttribute = gMacPibAssociationPermit_c;
boolFlag = TRUE;
pMsg->msgData.setReq.pibAttributeValue = &boolFlag;
ret = MSG_Send(NWK_MLME, pMsg);

/* This is a MLME-START.req command */
pMsg->msgType = gMlmeStartReq_c;

/* Create the Start request message data. */
pStartReq = &pMsg->msgData.startReq;
/* PAN ID - LSB, MSB. The predefinition shows a PAN ID of 0xAAAA. */
memcpy(pStartReq->panId, (void *)panId, 2);
/* Logical Channel - the default of 11 will be overridden */
pStartReq->logicalChannel = logicalChannel;

/* Beacon Order - 0xF = turn off Beacons */
pStartReq->BeaconOrder = 0x0F;
/* Superframe Order - 0xF = turn off Beacons */
pStartReq->superFrameOrder = 0x0F;
/* Be a PAN Coordinator */
pStartReq->panCoordinator = TRUE;
/* Don't use battery life extension */
pStartReq->batteryLifeExt = FALSE;
/* This is not a Realignment command */
pStartReq->coordRealignment = FALSE;
/* Don't use security */
pStartReq->securityEnable = FALSE;

/* Send the Start request to the MLME. */
if(MSG_Send(NWK_MLME, pMsg) == gSuccess_c)
{
```

```
        return errorNoError;
    }
  }
}
```
****

Code Snippet 12.9: You should be able to match up the Frame 2 Daintree Networks SNA capture text with the logic of the code you see here.

This should be ringing some bells. All we're doing here with the first NWK message is setting up the PAN Coordinator's short address in the MAC PIB. The second NWK message (another MAC PIB SET) we send sets up the allowance of association by the PAN Coordinator. The final message in the sequence starts the PAN Coordinator with all of the listed network attributes, which just happen to match up to the Frame 2 Daintree Networks SNA sniff.

The new PAN Coordinator will remain in the stateStartCoordinatorWaitConfirm state until a valid message of type gNwkStartCnf_c is returned. Of course, in Code Snippet 12.10, all along we are pretending that all of the return codes are positive

Code Snippet 12.10
****
```
    case stateStartCoordinatorWaitConfirm:
      /* Stay in this state until the Start confirm message
         arrives, and then goto the Listen state. */
      ret = App_WaitMsg(pMsgIn, gNwkStartCnf_c);
      if(ret == errorNoError)
      {
        state = stateListen;
      }
      break;
```
****
Code Snippet 12.10: Most of successful coding comes as a result of a positive attitude.

The new PAN Coordinator is now sitting and waiting for association requests in a state called stateListen. On the other side of the 802.15.4 world, the End Device's sun is rising.

The code is similar, but the ZigBee End Device has a totally different agenda as you can see in the End Device state list that follows in Code Snippet 12.11:

Code Snippet 12.11
****
```
enum
{
  stateInit,
  stateScanActiveStart,
  stateScanActiveWaitConfirm,
  stateAssociate,
  stateAssociateWaitConfirm,
  stateListen,
```

```
    stateTerminate
};
```
****
Code Snippet 12.11: Déjà vu, almost…

Since the PAN Coordinator and End Device are both MC13213-fed, the initialization process for the PAN Coordinator is identical to that of the End Device. You can also rest assured that the stateScanActiveStart code executed by the End Device is identical to that of the PAN Coordinator, with the only exception being the active scan argument passed by the End Device.

The events occurring within the stateScanActiveWaitConfirm state are worth taking a look at. So, please follow along with the stateScanActiveWaitConfirm code in Code Snippet 12.12:

Code Snippet 12.12

****
```
    case stateScanActiveWaitConfirm:
      /* Stay in this state until the Scan confirm message arrives, and then
goto the associate state or do a rescan in case of invalid short address.
      */

      /* ALWAYS free the Beacon frame contained in the Beacon notify
indication.*/
      rc = App_WaitMsg(pMsgIn, gNwkBeaconNotifyInd_c);
      if(rc == errorNoError)
      {
        MSG_Free(((nwkMessage_t *)pMsgIn)-msgData.BeaconNotifyInd.
pBufferRoot);
      }

      /* Handle the Scan Confirm message. */
      rc = App_WaitMsg(pMsgIn, gNwkScanCnf_c);
      if(rc == errorNoError)
      {
        rc = App_HandleScanActiveConfirm(pMsgIn);
        if(rc == errorNoError)
        {
          state = stateAssociate;
        }
        else
        {
          /* Restart scanning */
          App_WaitBusy( WAIT_INTERVAL_SEC );
          state = stateScanActiveStart;
        }
      }
      break;
```
****
Code Snippet 12.12: This is akin to sitting in a shuttle craft with Spock while he is performing a "short range" scan.

The gNwdBeaconNotifyInd_c message type must be received and processed as it must be removed from the message queue (that's a Freescale Semiconductor thing, not an 802.15.4 thing). Recall that indication primitives are used to notify layers of resultant primitive activity that may affect them.

When the active scan is complete, a confirmation is returned and handled by the App_HandleScanActiveConfirm function you see in Code Snippet 12.13:

Code Snippet 12.13

```
******************************************************************************
uint8_t App_HandleScanActiveConfirm(nwkMessage_t *pMsg)
{
  uint8_t panDescListSize   = pMsg->msgData.scanCnf.resultListSize;
  panDescriptor_t *pPanDesc = pMsg->msgData.scanCnf.resList.
pPanDescriptorList;
  uint8_t rc = errorNoScanResults;

  /* Check if the scan resulted in any coordinator responses. */
  if(panDescListSize != 0)
  {
    /* Initialize link quality to very poor. */
    uint8_t i, bestLinkQuality = 0;

    /* Check all PAN descriptors. */
    for(i=0; i<panDescListSize; i++, pPanDesc++)
    {
      /* Only attempt to associate if the coordinator
         accepts associations and is non-Beacon. */
      if( ( pPanDesc->superFrameSpec[1] & gSuperFrameSpecMsbAssocPermit_c) &&
          ((pPanDesc->superFrameSpec[0] & gSuperFrameSpecLsbBO_c) == 0x0F) )
      {
        /* Find the nearest coordinator using the link quality measure. */
        if(pPanDesc->linkQuality > bestLinkQuality)
        {
          /* Save the information of the coordinator candidate. If we
             find a better candidate, the information will be replaced. */
          memcpy(&coordInfo, pPanDesc, sizeof(panDescriptor_t));
          bestLinkQuality = pPanDesc->linkQuality;
          rc = errorNoError;
        }
      }
    }
  }

  /* ALWAYS free the PAN descriptor list */
  MSG_Free(pMsg->msgData.scanCnf.resList.pPanDescriptorList);

  return rc;
}
******************************************************************************
```
Code Snippet 12.13: End Devices can be programmed to be picky about which PAN they join.

The App_HandleScanActiveConfirm code parses a list of PAN Coordinator Beacon responses looking for the best link quality (highest radiation) figure of those in the list. This will be the closest or most powerful Coordinator in the End Device's POS (Personal Operating Space). If a suitable PAN Coordinator is detected, the application advances the state machine to the stateAssociate state. The End Device association process begins with this code sequence depicted in Code Snippet 12.14:

Code Snippet 12.14

```
********************************************************************************
    case stateAssociate:
      /* Associate to the PAN Coordinator */
      rc = App_SendAssociateRequest();
      if(rc == errorNoError)
        state = stateAssociateWaitConfirm;
      break;
********************************************************************************
```

Code Snippet 12.14: This is the one!

You should be getting the hang of the Freescale Semiconductor methods by now. Sniffer Capture 12.3 is the next frame in the Daintree Networks SNA sniff sequence associated with our fledgling network:

Sniffer Capture 12.3

```
********************************************************************************
Frame 3 (Length = 21 bytes)
       Time Stamp: 12:21:17.647
       Frame Length: 21 bytes
       Capture Length: 21 bytes
       Link Quality Indication: 152
IEEE 802.15.4
       Frame Control: 0xc823
                 .... .... .... .011  = Frame Type: Command (0x0003)
                 .... .... .... 0...  = Security Enabled: Disabled
                 .... .... ...0 ....  = Frame Pending: No more data
                 .... .... ..1. ....  = Acknowledgment Request: Acknowledgment
required
                 .... .... .0.. ....  = Intra PAN: Not within the PAN
                 .... ..00 0... ....  = Reserved
                 .... 10.. .... ....  = Destination Addressing Mode: Address
field contains a 16-bit short address (0x0002)
                 ..00 .... .... ....  = Reserved
                 11.. .... .... ....  = Source Addressing Mode: Address field
contains a 64-bit extended address (0x0003)
       Sequence Number: 238
       Destination PAN Identifier: 0xaaaa
       Destination Address: 0xcafe
       Source PAN Identifier: 0xffff
       Source Address: 0x9d2804b037c25000
       MAC Payload
```

```
                Command Frame Identifier = Association Request: (0x01)
                Capability Information: 0x80
                      .... ...0  = Alternate PAN Coordinator: Not capable of
becoming PAN Coordinator
                      .... ..0.  = Device Type: RFD
                      .... .0..  = Power Source: Not receiving power from
alternating current mains
                      .... 0...  = Receiver on when idle: Disables receiver
when idle
                      ..00 ....  = Reserved
                      .0.. ....  = Security Capability: Not capable of using
security suite
                      1... ....  = Allocate Address: Coordinator should
allocate short address
        Frame Check Sequence: Correct
```

```
0000:   23 c8 ee aa aa fe ca ff ff 00 50 c2 37 b0 04 28    #Hn**~J...PB70.(
0010:   9d 01 80 .. ..                                     .....
```
****

Sniffer Capture 12.3: The End Device is screaming "Associate me and gimme an address so I can play on your PAN."

The active scan information was parsed for the most eligible PAN Coordinator according to link quality. The lucky PAN Coordinator's information was pushed into a global PAN Coordinator buffer for use by the App_SendAssociateRequest function in Code Snippet 12.15:

Code Snippet 12.15
****
```
uint8_t App_SendAssociateRequest(void)
{
  mlmeMessage_t *pMsg;
  mlmeAssociateReq_t *pAssocReq;

  /* Allocate a message for the MLME message. */
  pMsg = MSG_AllocType(mlmeMessage_t);
  if(pMsg != NULL)
  {
    /* This is a MLME-ASSOCIATE.req command. */
    pMsg->msgType = gMlmeAssociateReq_c;

    /* Create the Associate request message data. */
    pAssocReq = &pMsg->msgData.associateReq;

    /* Use the coordinator info we got from the Active Scan. */
    memcpy(pAssocReq->coordAddress, coordInfo.coordAddress, 8);
    memcpy(pAssocReq->coordPanId,   coordInfo.coordPanId, 2);
    pAssocReq->coordAddrMode     = coordInfo.coordAddrMode;
    pAssocReq->logicalChannel    = coordInfo.logicalChannel;
    pAssocReq->securityEnable    = FALSE;
    /* We want the coordinator to assign a short address to us. */
    pAssocReq->capabilityInfo    = gCapInfoAllocAddr_c;
```

```
    /* Send the Associate Request to the MLME. */
    if(MSG_Send(NWK_MLME, pMsg) == gSuccess_c)
    {
      return errorNoError;
    }
  }
}
```

Code Snippet 12.15: Remember, data doesn't just fall out of the sky. Something or someone always generates data. You'll find that if you catch the right bytes, you can use them to fill holes such as what channel are we on or what was that Coordinator's address.

The capabilityInfo changes included only a single bit, which asked to have the PAN Coordinator assign the End Device address. This change was indicated in the Frame 3 Daintree Networks SNA sniff. Let's fast forward, as we've seen plenty of acknowledgment frames and we should be familiar with the association sequence. Frame 4 of the Daintree Networks SNA sniff sequence is the expected acknowledgment frame, which is followed by a Data Request frame in Frame 5 and another acknowledgment in Frame 6. The sniff data in Frame 7 (Sniffer Capture 12.4) is what the End Device is waiting for:

Sniffer Capture 12.4

```
Frame 7 (Length = 27 bytes)
        Time Stamp: 12:21:18.146
        Frame Length: 27 bytes
        Capture Length: 27 bytes
        Link Quality Indication: 136
IEEE 802.15.4
        Frame Control: 0xcc63
                .... .... .... .011 = Frame Type: Command (0x0003)
                .... .... .... 0... = Security Enabled: Disabled
                .... .... ...0 .... = Frame Pending: No more data
                .... .... ..1. .... = Acknowledgment Request: Acknowledgment
required
                .... .... .1.. .... = Intra PAN: Within the PAN
                .... ..00 0... .... = Reserved
                .... 11.. .... .... = Destination Addressing Mode: Address
field contains a 64-bit extended address (0x0003)
                ..00 .... .... .... = Reserved
                11.. .... .... .... = Source Addressing Mode: Address field
contains a 64-bit extended address (0x0003)
        Sequence Number: 13
        Destination PAN Identifier: 0xaaaa
        Destination Address: 0x9d2804b037c25000
        Source Address: 0x820b04b037c25000
        MAC Payload
                Command Frame Identifier = Association Response: (0x02)
                Short Address: 0x0001
                Association Status: Association Successful (0x00)
        Frame Check Sequence: Correct
```

```
0000:    63 cc 0d aa aa 00 50 c2 37 b0 04 28 9d 00 50 c2    cL.**.PB70.(..PB
0010:    37 b0 04 0b 82 02 01 00 00 .. ..                   70........
```
****
Sniffer Capture 12.4: Look at all of the "data" we collect including our new short address.

The Frame 7 sniff satisfies the obstacles laid down by the stateAssociateWaitConfirm state and the End Device application advances into the stateListen state in the code you see in Code Snippet 12.16:

Code Snippet 12.16
****
```
    case stateAssociateWaitConfirm:
      /* Stay in this state until the Associate confirm message
         arrives, and then goto the Listen state. */
      rc = App_WaitMsg(pMsgIn, gNwkAssociateCnf_c);
      if(rc == errorNoError)
      {
        /* Check for coordinator at full capacity error */
        if (App_HandleAssociateConfirm(pMsgIn) == gSuccess_c)
        {
          state = stateListen;
        }
        else
        {
          /* Restart scanning */
          App_WaitBusy(WAIT_INTERVAL_SEC);
          state = stateScanActiveStart;
        }
      }
      break;
```
****
Code Snippet 12.16: Again, we're taking a positive attitude that the code won't fail out here and start a rescan process.

If something goes wrong in the association sequence, the End Device will attempt to associate again, following a rescan of the ether.

OK…That does it for getting the PAN established and getting an End Device to join it. What you're really here for is the sending and receiving party. So, let's put some code together that matches the Daintree Networks SNA sniff in Sniffer Capture 12.5:

Sniffer Capture 12.5
****
```
Frame 2300 (Length = 12 bytes)
      Time Stamp: 12:26:16.271
      Frame Length: 12 bytes
      Capture Length: 12 bytes
      Link Quality Indication: 152
IEEE 802.15.4
      Frame Control: 0x8861
```

```
                  .... .... .... .001  = Frame Type: Data (0x0001)
                  .... .... .... 0...  = Security Enabled: Disabled
                  .... .... ...0 ....  = Frame Pending: No more data
                  .... .... ..1. ....  = Acknowledgment Request: Acknowledgment
required
                  .... .... .1.. ....  = Intra PAN: Within the PAN
                  .... ..00 0... ....  = Reserved
                  .... 10.. .... ....  = Destination Addressing Mode: Address
field contains a 16-bit short address (0x0002)
                  ..00 .... .... ....  = Reserved
                  10.. .... .... ....  = Source Addressing Mode: Address field
contains a 16-bit short address (0x0002)
        Sequence Number: 128
        Destination PAN Identifier: 0xaaaa
        Destination Address: 0x0001
        Source Address: 0xcafe
        Frame Check Sequence: Correct
MAC Payload: 01

0000:    61 88 80 aa aa 01 00 fe ca 01 .. ..              a..**..~J...
*************************************************************************
```

Sniffer Capture 12.5: Can you pick out the payload? Who's sending this? If you couldn't answer either question, get some help from your baby sister.

OK…This little bugger is coming from the PAN Coordinator and is aimed at the End Device. So, let's get our data in place first. We do just that in Code Snippet 12.17:

Code Snippet 12.17
```
*************************************************************************
    deviceAddress = 0x0001;
    dataBuffer[0] = 0x01;
    msduLength = 1;
*************************************************************************
```
Code Snippet 12.17: None of this should be foreign to you by now.

The next step involves creating an MCPS-Data Request message containing the data we want to send. I've inserted the corresponding Daintree Networks SNA sniff data within the lines of code that follow in Code Snippet 12.18:

Code Snippet 12.18
```
*************************************************************************
    //Frame Type: Data (0x0001)
    pPacket->msgType = gMcpsDataReq_c;

    // Copy data to be sent to packet
    memcpy(pPacket->msgData.dataReq.msdu, (void *)dataBuffer,msduLength);

    //Destination Address: 0x0001
    pPacket->msgData.dataReq.dstAddr[0] = deviceAddress;
    pPacket->msgData.dataReq.dstAddr[1] = 0;
```

```
//Source Address: 0xcafe
memcpy(pPacket->msgData.dataReq.srcAddr, (void *)shortAddress, 2);

//Destination PAN Identifier: 0xaaaa
memcpy(pPacket->msgData.dataReq.dstPanId, (void *)panId, 2);

// const uint8_t panId[2] = { 0xAA, 0xAA };
memcpy(pPacket->msgData.dataReq.srcPanId, (void *)panId, 2);

//Destination Addressing Mode
//Address field contains a 16-bit short address (0x0002)
pPacket->msgData.dataReq.dstAddrMode = gAddrModeShort_c;

//Source Addressing Mode
//Address field contains a 16-bit short address (0x0002)
pPacket->msgData.dataReq.srcAddrMode = gAddrModeShort_c;

//msduLength = 1;
pPacket->msgData.dataReq.msduLength = msduLength;

//Acknowledgment Request: Acknowledgment required
pPacket->msgData.dataReq.txOptions = gTxOptsAck_c | gTxOptsIndirect_c;

//msduHandle is arbitrary but associated with this message
pPacket->msgData.dataReq.msduHandle = msduHandle++;

// Send the Data Request to the MCPS
NR MSG_Send(NWK_MCPS, pPacket);
```
****
Code Snippet 12.18: This is how I've been trying to get you to think.

Wanna send it in the other direction? No worries. Just turn the addresses around and send it from the End Device. The basic operational code in the End Device is identical to that in the PAN Coordinator. I hard-coded the data byte in this example. However, that data byte could be any data that you want to send. It could come from an analog-to-digital converter read of a sensor or a simple switch closure. The key is packaging it correctly and being aware of the proper primitives to employ in the transmit operation.

What we don't see in the Daintree Networks SNA sniff sequence is the reception and subsequent processing of the received data. The whole idea of how you process incoming data depends on what you do with it when you pull it out of the radio's buffer. The C source in Code Snippet 12.19 is the Freescale Semiconductor way of organizing some of the 802.15.4 frame data contents:

Code Snippet 12.19
****
```
/* Pointer for storing the messages from MLME, MCPS, and ASP. */
void *pMsgIn;
```

```
void App_HandleMessage(mcpsToNwkMessage_t *pMsgIn)
{
  uint8_t val = *(pMsgIn->msgData.dataInd.msdu);
  uint8_t len = *(pMsgIn->msgData.dataInd.msduLength);
  uint8_t hex = *(pMsgIn->msgData.dataInd.mpduLinkQuality);
}
```
****
Code Snippet 12.19: Just take a deep breath and read each line left to right. It will make sense.

You can make it as simple or as complex as your mind will allow you to. If you make it hard, you're on your own as Freescale Semiconductor's ZigBee methodology has really eliminated the complexity.

Unfortunately, I got the new release of Freescale Semiconductor's BeeKit a bit too late to expound upon its features in the pages of this book. However, I can tell you that the Freescale Semiconductor BeeKit application will work with the Freescale Semiconductor IEEE 802.15.4/ZigBee components we discussed earlier. The idea behind the BeeKit is to allow you to define your IEEE 802.15.4 or ZigBee application in the BeeKit template environment. BeeKit then takes all of the stuff you fed it and assembles a project that you can export from the BeeKit environment. The exported project is then loaded into the Freescale Semiconductor CodeWarrior IDE, where it can be edited and compiled.

About Freescale Semiconductor

You knew them as Motorola. However, Freescale Semiconductor is a totally different animal. This isn't my first Freescale Semiconductor diatribe, as I've done a few Freescale Semiconductor magazine columns. Freescale Semiconductor's efforts in the IEEE 802.15.4/ZigBee arena are obvious as within the pages of this book you've already seen Freescale Semiconductor IEEE 802.15.4-compliant/ZigBee-ready radios used in other manufacturers' IEEE 802.15.4/ZigBee products.

You also knew them as the Supremes, Diana Ross, Mary Wilson and Florence Ballard. Here are the hits in their order of appearance:

- Where Did Our Love Go
- Baby Love
- Come See About Me
- Stop! In the Name of Love

I hope your sister isn't reading this right now. OK…So you aren't real good with R&B (rhythm and blues) groups. How about this? What very famous country artist loved Billy Joel and what Billy Joel song did he perform regularly?

Panasonic

You've had a pretty thorough introduction to the Freescale Semiconductor way of doing business with IEEE 802.15.4 applications. In this chapter, we will step through the procedures that I had to execute to get my pair of Panasonic PAN802154HAR00 IEEE 802.15.4-compliant/ZigBee-ready modules to talk intelligently to each other.

The Panasonic PAN802154HAR00 is based on Freescale Semiconductor technology. Looking at Figure 13.1, you can see that the Panasonic PAN802154HAR00 is a combination of a Freescale Semiconductor MC9S08GT60CFB microcontroller and a Freescale Semiconductor MC13193 IEEE 802.15.4-compliant transceiver.

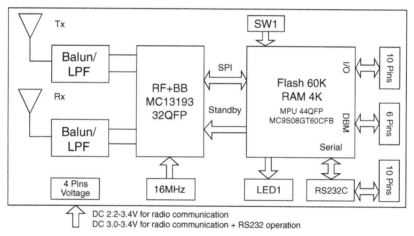

Figure 13.1: Note the twin antennae. I can see lots of you putting this little bugger to real-world use, as it is very compact and easy to work with. I had a blast with these radios.

The only frills included with the Panasonic PAN802154HAR00 hardware are a user-available LED and pushbutton switch. To accommodate RS-232-capable devices that use the official RS-232 voltages, the Panasonic PAN802154HAR00 is also equipped with a true RS-232 serial port, which is pinned out to a 10-pin male header as shown in Figure 13.2.

All you need to do to ready the Panasonic PAN802154HAR00 module hardware is to build up the serial cable (if you have RS-232 in your IEEE 802.15.4 or ZigBee application) and provide power to the module. I made up an RS-232 cable for my Panasonic PAN802154HAR00 modules and I fabricated a removable power connector like the one you see in Photo 13.1.

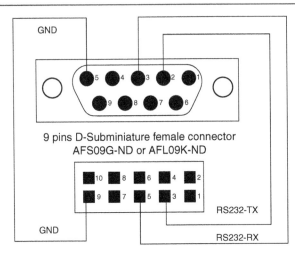

Figure 13.2: Don't worry about following the schematic here, as all you have to do is point both of the connector's pins up just as you see them in the figure and stamp in a 10-position ribbon cable between them. It's automatic.

Photo 13.1: This is nothing more than a standard wall-wart power jack. I chopped off a piece of a double-row header and attached it to the power jack across two pins on each side. Pins 1 and 3 (right) are grounded while pins 2 and 4 (left) are positive voltage (+3.3 VDC). And yes, each and every one of these in that 1,000-piece bag is marked "MADE IN CHINA." Amazing, isn't it?

The method to my madness with the power jack can be seen in the next photo, Photo 13.2, which is a side shot of the Panasonic PAN802154HAR00 with the power jack assembly mounted.

The RS-232 cable is the final optional attachment that needs to be made to make the Panasonic PAN802154HAR00 module battlefield-ready. I flipped the Panasonic PAN802154HAR00 onto its back to give you a Panasonic-ramic view in Photo 13.3.

OK…Now that you've been indoctrinated, a low-level reconnaissance photo of the Panasonic PAN802154HAR00 is captured in Photo 13.4.

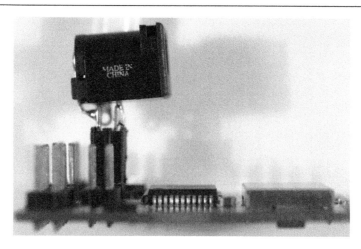

Photo 13.2: The PEmicro USB Multilink Interface pins (BDM pins) are visible directly in front of the power jack pins. Two sets of 10-pin headers behind the power jack and BDM (Background Debug Mode) pins take on the RS-232 port and provide general-purpose I/O access.

Photo 13.3: The TX and RX antennae are just under their silkscreen names. The radio is under the shielding. Imagine that! That's an RS-232 cable at the bottom left. I'll bet you can tell me what the IC to its right is doing.

The overall concept of the Panasonic PAN802154HAR00 is very simple. Fortunately, if you follow the rules, the firmware that drives the Panasonic PAN802154HAR00 is just as easy to comprehend. I didn't follow the rules, which is a good thing in this case. I can show you what not to do.

You'll need a couple of other things to be successful with the Panasonic PAN802154HAR00. The Freescale Semiconductor IEEE 802.15.4/ZigBee development tool chain is a must. And,

Photo 13.4: This little puppy measures in at only 1.375" × 1.325". You can put metal screws in every hole except the holes at the TX and RX silkscreen legends, which must be

don't count on getting far with the "special" editions of the CodeWarrior compiler. You'll need to open the wallet for the full version to open up the full ZigBee capabilities of the Panasonic PAN802154HAR00 hardware. It is possible to get by with the low-code-count compiler packages if you're only doing IEEE 802.15.4 or Freescale Semiconductor SMAC stuff.

The other must-have is the nifty tool in Photo 13.5, the PEmicro USB Multilink Interface BDM module. You can use the PEmicro BDM in conjunction with the HI-WAVE application and the CodeWarrior IDE without having to load bootloader firmware and use personal computer-based bootloader applications to program the Panasonic PAN802154HAR00 microcontroller's Flash. The module in Photo 13.5 will also act as a Ford tractor and pull your butt out of the ditch when you erase too much in the Panasonic PAN802154HAR00 microcontroller's program Flash.

Photo 13.5: This dongle has saved my butt many times during the writing of the Freescale Semiconductor-based chapters of this book. It is reliable and easy to put into play even when you're in the ditch. I like the way the Panasonic PAN802154HAR00 application note puts it: "For serious application the BDM tool is recommended."

Here's how it all went down with the Panasonic PAN802154HAR00 modules. After applying power to both Panasonic PAN802154HAR00 modules, I turned on the Daintree Networks SNA application and attempted to sniff out anything that may have been coming out of either Panasonic PAN802154HAR00 module. No luck.

The Panasonic PAN802154HAR00 modules are very similar to the Freescale Semiconductor SARD development platform. You've been reading about IEEE 802.15.4 and ZigBee theory for a while now (and I've been writing about it for a while now). So, I want to show you something that could be very useful to you right out of the box.

Let's begin by putting some PAN Coordinator code into one of the Panasonic PAN-802154HAR00 modules. If one were to actually take some time and look, he or she would find a program file called HIWAVE.EXE in the CodeWarrior prog directory. I clicked on that bugger and what you see in Screen Capture 13.1 is what I saw.

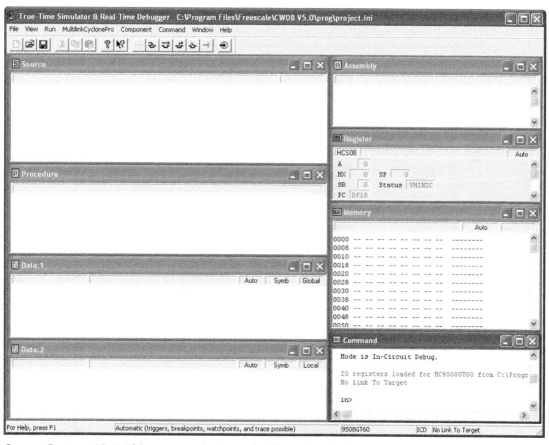

Screen Capture 13.1: This program is also called from the CodeWarrior IDE when you enter debug mode. We only need it to interact with the PEmicro BDM.

The ball begins to roll right here. I'm in the throes of establishing the initial connection to the BDM and it is essential that I select HCS08 in relation to the obvious for the connection in Screen Capture 13.2.

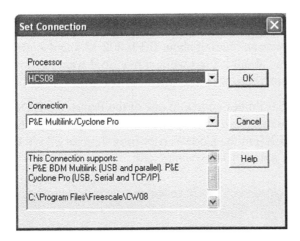

Screen Capture 13.2: I went the wrong way the first time and chose HC08 instead of HCS08. Trust me, you'll get a different path with HC08.

Finally, I got the window you see in Screen Capture 13.3. The good news is that the BDM has been detected and recognized by the HI-WAVE application.

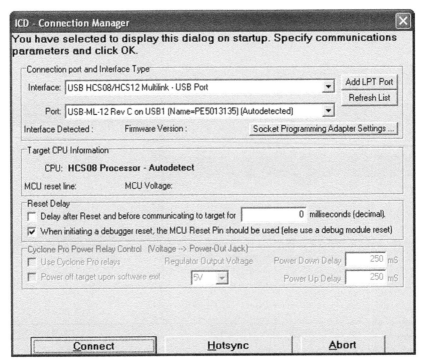

Screen Capture 13.3: I got a blank-filled window like this when I went down the dead-end wrong path with my incorrect selection of the HC08. There was nothing I could enter to get any satisfaction from the blank-filled window.

The "Target Ready" in Screen Capture 13.4 brought a sigh of relief. I am now poised on the verge of success.

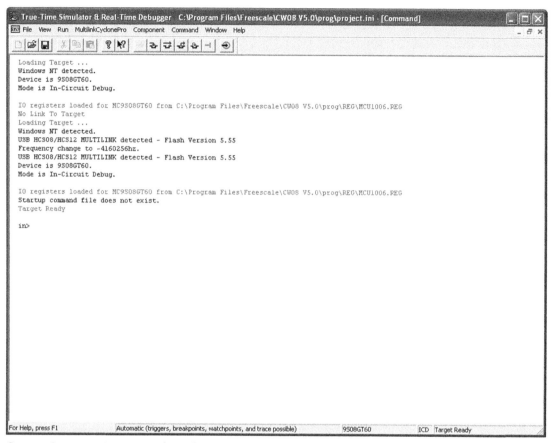

Screen Capture 13.4: Even the microcontroller on the Panasonic PAN802154HAR00 was detected correctly.

The Panasonic web site offered up a couple of Jiffy-Cornbread-Mix ZigBee applications. For those of you that are not familiar with Jiffy Cornbread Mix, it is a sweet cornbread mix used by many young and single Southern women on unsuspecting young and single Southern men. Once the man is clinched, the Jiffy mix gets replaced by the real thing. That's sorta like putting the Modem Coordinator code into a blank Panasonic PAN802154HAR00 module. The window in Screen Capture 13.5 is a result of selecting the Load Application option from the HI-WAVE File dropdown menu.

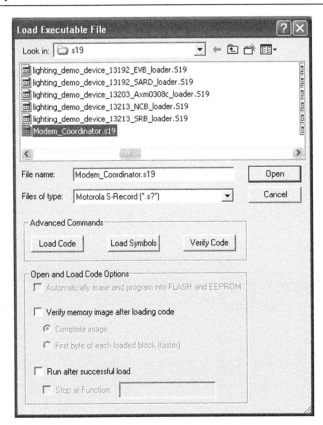

Screen Capture 13.5: The secret to getting your desired hex file into this window is to place it into the s19 directory of the Freescale Semiconductor Test Tool directory.

After clicking on the Open button in Screen Capture 13.5, I was presented with the choices in Screen Capture 13.6.

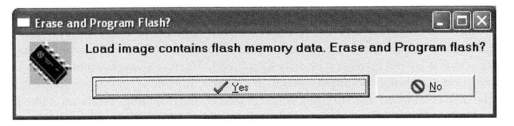

Screen Capture 13.6: Wanna give me a gun so I can shoot myself in the foot, or do I just say YES here?

Duh huh, as Bubba would say. I punched in a YES in Screen Capture 13.6 and then proceeded to repeat the entire process we just walked through to put the End Device code into the second Panasonic PAN802154HAR00 module in my possession. The Jiffy Cornbread Mix for the End Device is shown in Screen Capture 13.7.

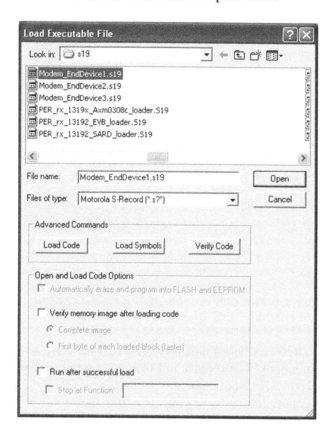

Screen Capture 13.7: As you can see here, we can open up to three boxes of End Device Jiffy Cornbread Mix.

Things are good. I fired up the Panasonic PAN802154HAR00 PAN Coordinator and received a Beacon Request according to Daintree Networks SNA. That means that, more than likely, the PAN Coordinator has scanned the channel and established a PAN there. The Panasonic PAN802154HAR00 documentation indicated that I may want to be on channel 11 when I fired up the Panasonic PAN802154HAR00 PAN Coordinator.

To make any sense of the application, you'll need to know the official 64-bit IEEE 802.15.4 addresses of each node. The easy way to show you that information is to simply show you the association response capture, which follows:

```
*******************************************************************************
Frame 24 (Length = 27 bytes)
        Time Stamp: 16:18:34.666
        Frame Length: 27 bytes
        Capture Length: 27 bytes
```

```
            Link Quality Indication: 80
IEEE 802.15.4
        Frame Control: 0xcc63
                .... .... .... .011  = Frame Type: Command (0x0003)
                .... .... .... 0...  = Security Enabled: Disabled
                .... .... ...0 ....  = Frame Pending: No more data
                .... .... ..1. ....  = Acknowledgment Request: Acknowledgment
required
                .... .... .1.. ....  = Intra PAN: Within the PAN
                .... ..00 0... ....  = Reserved
                .... 11.. .... ....  = Destination Addressing Mode: Address
field contains a 64-bit extended address (0x0003)
                ..00 .... .... ....  = Reserved
                11.. .... .... ....  = Source Addressing Mode: Address field
contains a 64-bit extended address (0x0003)
        Sequence Number: 0
        Destination PAN Identifier: 0x1110
        Destination Address: 0x2726252423222120
        Source Address: 0x1716151413121110
        MAC Payload
                Command Frame Identifier = Association Response: (0x02)
                Short Address: 0x796f
                Association Status: Association Successful (0x00)
        Frame Check Sequence: Correct

0000:   63 cc 00 10 11 20 21 22 23 24 25 26 27 10 11 12    cL... !"#$%&'...
0010:   13 14 15 16 17 02 6f 79 00 .. ..                   .....oy...
********************************************************************************
```

I get the feeling that the IEEE 802.15.4 addresses here aren't very official. That's OK, as we're just playing around in the pages of a book. The applications that are currently running on both the Panasonic PAN802154HAR00 PAN Coordinator and the Panasonic PAN-802154HAR00 End Device take in a specially formatted hex message from the serial port and ship it across the IEEE 802.15.4 link to the remote node. I'm using a personal computer as the serial input device but, as you will quickly see, the source serial device can easily be your favorite microcontroller running a very simple characters-out-of-the-serial port routine that transfers whatever data you want to transmit to any active node in the PAN.

Figure 13.3 is how the RS-232 packet must be formatted. Now you can see why it is important for you to know the 64-bit address of the node you wish to contact. We see it here physically via the Daintree Networks SNA sniff. However, you already know that you can obtain the address you need programmatically as well.

Tera Term Pro allows me to send a binary stream out of the personal computer's serial port. So, all I need to do is build up a message using the format outlined in Figure 13.3. No worries. Check out the hex message I created with HHD's free hex editor in Screen Capture 13.8.

I fired off the message in Screen Capture 13.8 at 38400 bps. The Panasonic PAN-802154HAR00 Coordinator's UART picked it up and turned it into IEEE 802.15.4 RF, resulting in the IEEE 802.15.4 frame you see in Sniffer Capture 13.1.

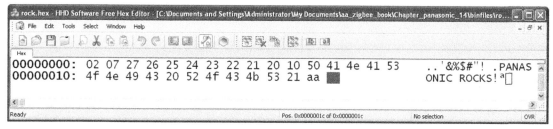

Screen Capture 13.8: Sixteen bytes of payload data is ready to go.

SOP	CMD	Dest MAC Addr	Len	Payload	EOP
0x02	0x07	0x2726252423222120	4	0xA 0xB 0xC 0xD	0xAA

Figure 13.3: RS-232 packet format. This is pretty straightforward. This should be no problem to generate using a microcontroller and its UART.

Sniffer Capture 13.1

****

```
Frame 302 (Length = 53 bytes)
        Time Stamp: 16:19:03.736
        Frame Length: 53 bytes
        Capture Length: 53 bytes
        Link Quality Indication: 116
IEEE 802.15.4
        Frame Control: 0x8861
                .... .... .... .001  = Frame Type: Data (0x0001)
                .... .... .... 0...  = Security Enabled: Disabled
                .... .... ...0 ....  = Frame Pending: No more data
                .... .... ..1. ....  = Acknowledgment Request: Acknowledgment
required
                .... .... .1.. ....  = Intra PAN: Within the PAN
                .... ..00 0... ....  = Reserved
                .... 10.. .... ....  = Destination Addressing Mode: Address
field contains a 16-bit short address (0x0002)
                ..00 .... .... ....  = Reserved
                10.. .... .... ....  = Source Addressing Mode: Address field
contains a 16-bit short address (0x0002)
        Sequence Number: 1
        Destination PAN Identifier: 0x1110
        Destination Address: 0x796f
        Source Address: 0x0000
        Frame Check Sequence: Correct
ZigBee NWK
        Frame Control: 0x0004
                .... .... .... ..00  = Frame Type: NWK Data (0x00)
                .... .... ..00 01..  = Protocol Version (0x01)
```

```
                .... .... 00.. ....    = Discover Route: Suppress route
discovery (0x00)
                .... ...0 .... ....    = Reserved
                .... ..0. .... ....    = Security: Disabled
                0000 00.. .... ....    = Reserved
        Destination Address: 0x796f
        Source Address: 0x0000
        Radius = 7
        Sequence Number = 1
NWK Payload: 40:0c:01:05:0f:0b:21:00:19:17:16:15:14:13:12:11:10:10:50:41:4e:
41:53:4f:4e:49:43:20:52:4f:43:4b:53:21


0000:   61 88 01 10 11 6f 79 00 00 04 00 6f 79 00 00 07    a....oy....oy...
0010:   01 40 0c 01 05 0f 0b 21 00 19 17 16 15 14 13 12    .@.....!........
0020:   11 10 10 50 41 4e 41 53 4f 4e 49 43 20 52 4f 43    ...PANASONIC ROC
0030:   4b 53 21 .. ..                                     KS!..
```
****

Sniffer Capture 13.1: The RS-232 wrapper is removed by the application and the received data is in the same format as the transmitted data. Therefore, the receiving application can simply parse into the message to pick up the payload.

The transmitting station actually replaces the 64-bit destination with its 64-bit address in the message body. Thus, as the received message is in the format indicated by Figure 13.3, the receiving station can simply parse the incoming message, retrieve the payload and retrieve the sender's address. Pretty danged cool!

About Panasonic

The Panasonic PAN802154HAR00 was suggested to me by a ZigBee programmer in South Florida. He really liked the Panasonic PAN802154HAR00 and so do I. There are many things to like about the Panasonic PAN802154HAR00 and being able to leverage your recently acquired Freescale Semiconductor knowledge against the Panasonic PAN802154HAR00 is one of them. Robert Nguyen also answers the phone and responds to email. During the course of talking to Robert, I really messed up his last name in an email address snafu. Robert, expecting to hear from me, probably figured I was a bit flakey as I hadn't responded to his emails. As it turned out, not hearing from Robert prompted me to take another look at his original email. I saw the error of my ways and corrected the email address. Robert immediately answered and was still willing to help me get the Panasonic PAN802154HAR00 word out. That's spelled N-g-u-y-e-n.

Hopefully, you answered the Motown questions correctly, or at least your sister has given the book back to you by now. The Billy Joel song is "Shameless." Do you have the country artist's name yet? How about Garth Brooks? Enough said.

Let's keep it country. Who is Bill Monroe?

# CHAPTER 14

DLP Design

Freescale Semiconductor's IEEE 802.15.4-compliant/ZigBee-ready radios are seemingly everywhere. Don Powrie at DLP Design has based a series of DLP Design IEEE 802.15.4-compliant/ZigBee-ready devices on the Freescale Semiconductor MC13193/MC9S08GT60 combination.

The differentiating factor of DLP Design's set of IEEE 802.15.4/ZigBee products is the inclusion of a proprietary application called SIPP that rides over the top of Freescale Semiconductor's SMAC. The SIPP firmware comes preprogrammed and ready to run in all of the DLP Design IEEE 802.15.4-compliant/ZigBee-ready radios. SIPP allows access to the DLP Design transceiver's functions by way of simple serial calls. The SIPP-equipped DLP-RF1-Z can be used in conjunction with other DLP-RF1-Z/DLP-RF2-Z transceivers, as well as other MC13193-based transceivers, to form point-to-point and star networks.

If SIPP is not your thing, you can use the DLP Design radios as raw Freescale Semiconductor IEEE 802.15.4/ZigBee platforms. Each DLP Design IEEE 802.15.4/ZigBee radio is equipped with a 6-pin BDM connector that can accommodate development tools such as the PEmicro USB Multilink Interface you're familiar with.

The first DLP Design IEEE 802.15.4-compliant/ZigBee-ready module we will talk about is the DLP-RF1-Z-Z, which you can marvel at in Photo 14.1.

The preprogrammed SIPP firmware allows the DLP-RF1-Z-Z to be used in point-to-point and star network configurations consisting of other DLP-RF1-Z-Z and/or DLP-RF2-Z modules. When used with the SIPP firmware, both the DLP-RF1-Z and the DLP-RF2-Z can serve as host/system controllers. The DLP-RF1-Z acting as a host requires a USB-capable desktop personal computer, embedded single-board computer, PDA or laptop running Windows, Linux, or Mac PC. If your application requires a microcontroller instead, the DLP-RF2-Z is your choice. All that is required to interface to the microcontroller is a 3-wire serial interface consisting of TX, RX, and ground connections. However, the DLP-RF2-Z can operate without a host processor. The SIPP firmware within the DLP-RF2-Z can be used to gain access to the MC9S08GT60's port pins for basic digital I/O.

Each transceiver running the SIPP firmware has a unique 16-bit ID, which is used in the transmission and reception of packets. Every SIPP data packet contains the number of bytes in the packet, the destination transceiver ID, the source transceiver ID, and a command byte as shown in Figure 14.1.

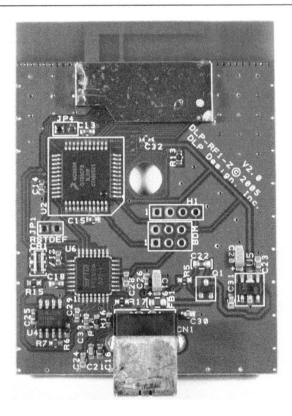

Photo 14.1: This is the USB version of the DLP Design IEEE 802.15.4-compliant/ZigBee-ready radio line. The DLP-RF1-Z-Z is intended to be tied to a personal computer.

Byte	Description	Comments
0	Number of bytes in the packet following byte 0: 5-124	Each packet must contain (as a minimum) the number of bytes, a destination ID, Source ID and a command byte.
1	Destination ID MSByte ID Range: 1-65535*	ID:1 default for new DLP-RF1 transceivers ID:2 default for new DLP-RF2 transceivers ID:0 reserved for broadcast to all transceivers
2	Destination ID LSByte	
3	Source ID MSByte Range: 1-65535	
4	Source ID LSByte	
5	Command Byte Command Range: 0xA0-0xDF	Both Command Packets and Reply Packets. Every packet must have a command byte.
6	Data Byte(s)	0-119 bytes of data are allowed in the packet.

Figure 14.1: Each SIPP packet must consist of at least five of the seven bytes shown in this figure.

The DLP-RF1-Z does not require that the network/application designer have any IEEE 802.15.4 or ZigBee knowledge (like you now possess). The reason for that is partially embedded within the partial command list I've posted in Figure 14.2.

Cmd	Packet Recipient	Description	Expected Reply Cmd
0xA0	MC9S08GT60	Ping (no data)	0xC0
0xA1	MC9S08GT60	Set Transmit Power Level 1 Data Byte; Range: 0-15	0xC0
0xA2	MC9S08GT60	Set Transceiver Channer 1 Data Byte; Range: 0-15	0xC0
0xA3	MC9S08GT60	Set RF2 Baud Rate (RF2 only)	0xC0
0xA4	MC9S08GT60	Release immediately to Sleep (DLP-RF2 only) (no data)	0xC0
0xA5	MC9S08GT60	Measure energy on all channels (no data)	0xC3
0xA6	MC9S08GT60	Return all packets received to host (Packet Watch Mode—no data)	0xC0
0xA7	MC9S08GT60	Return only packets with correct ID to host (Default mode) (no data)	0xC0
0xA8	MC9S08GT60	Read EEPROM 1 Data Byte; Address: 0-31	0xC4
0xA9	MC9S08GT60	Write EEPROM and update checksum 2 Data Bytes; Address: 0-30; Data: 0-255	0xC5
0xAA	MC9S08GT60	Read I/O pin, 1 Data Byte: Port: 0-7 (RF2 only)	0xC6
0xAB	MC9S08GT60	Set I/O pin direction, 2 Data Bytes: Port: 0-7, Direction: 1=Out, 0=In (RF2 only)	0xC7
0xAC	MC9S08GT60	Set/Clear I/O pin, 2 Data Bytes: Port: 0-7, State: 0/1 (RF2 only)	0xC8
0xAD	MC9S08GT60	Setup A/D 2 Data Bytes: Port: 0-6, Mode: 0=Off, 1=On (RF2 only)	0xC9
0xAE	MC9S08GT60	Read A/D, 1 Data Byte: Channel: 0-6 (RF2 only)	0xCA
0xAF	MC9S08GT60	Read VBAT (no data) (RF2 only)	0xCB
0xB6	MC9S08GT60	Request Board Type (DLP-RF1, RF2), ROM and RFIC versions (no data)	0xCD
0xB7	MC9S08GT60	Return Board ID (not available through RF transceiver, physical connection only) (destination ID ignored) (no data)	0xCF

Figure 14.2: There are a couple of primitive related commands here, as well as some general-purpose I/O commands. The only difference is that you really don't have to know what a primitive is or how it works to control things with the DLP Design radios and SIPP firmware.

The DLP-RF1-Z uses a USB interface design that is backed by VCP (Virtual COM Port) drivers that reside on the personal computer computing platform. Once the VCP drivers are installed, the DLP-RF1-Z appears to the host PC as a standard RS-232 COM port. The VCP drivers intercept the application's data packets on their way to the RS-232 COM port and reroute them to the USB interface. Incoming USB traffic is converted to RS-232-compatible data by an FTDI USB-to-RS-232 IC and transferred directly to the DLP-RF1-Z microcontroller's UART.

The DLP-RF1-Z's alter ego is the DLP-RF2-Z you see in Photo 14.2. The DLP-RF2-Z is designed to interface to an external microcontroller if your application requires that. Otherwise, the DLP-RF2-Z's general-purpose I/O can be manipulated remotely using only the DLP-RF2-Z's on-chip SIPP firmware.

Photo 14.2: What you don't see here is a 20-pin dual-row header on the other side of the DLP-RF2-Z. The actual MC13193 is under the hood to the right of the shot.

I'm sure that you noticed that there are some SIPP commands that only the DLP-RF2-Z can handle. To that end, I've posted the remainder of the SIPP command set in Figure 14.3.

0xAC	MC9S08GT60	Set/Clear I/O pin, 2 Data Bytes: Port: 0-8 (Port 8 is PTC0), State: 0/1	0xC8
0xAD	MC9S08GT60	Setup A/D, 2 Data Bytes: Port: 0-6, Mode: 0=Off, 1=On	0xC9
0xAE	MC9S08GT60	Read A/D, 1 Data Byte: Channel: 0-6	0xCA
0xAF	MC9S08GT60	Read VBAT (no data)	0xCB
0xB6	MC9S08GT60	Request Board Type (DP-RF1, RF2-Z), ROM and RFIC versions (no data)	0xCD
0xB7	MC9S08GT60	Return Board ID (not available through RF transceiver, physical connection only) (destination ID ignored) (no data)	0xCF
0xB8	MC9S08GT60	Pulse high/low with delay while high (For DLP-RF2-ZRELAY only; additional hardware required) 2 Data Bytes: Relay Number: 1/2, State: 0/1 (RST/SET)	0xC0
0xB9	MC9S08GT60	Read Temperature and Humidity (DLP-RF2-Z only; additional hardware required) (no data) Refer to RFTestAP source code for data processing details	0xCC
0xC0	Serial / USB	Generic Reply or "ACK" for selected non-broadcast commands	

0xC1	Serial / USB	Check-in from DLP-RF2-Z due to monitored port pin input change. 2 Data Bytes: Current I/O pin state (A6, B6:0), Bit-field with bits set for the port pins that changed state (A6, B6:0)	
0xC2	Serial / USB	Check-in from DLP-RF2-Z due to wake from sleep (no data)	
0xC3	Serial / USB	Measured energy data, 16 Data ByTes: Channel 0 – channel 15 energy levels, Refer to RFTestAp source code for data processing details	
0xC4	Serial / USB	EEPROM read reply, 1 Data Byte: EEPROM Read data	
0xC5	Serial / USB	Write EEPROM reply (no data)	
0xC6	Serial / USB	Read I/O pin reply, 1 Data Byte: pin state	
0xC7	Serial / USB	Set direction reply (no data)	
0xC8	Serial / USB	Set/Clear I/O pin reply (no data)	
0xC9	Serial / USB	Setup A/D reply (no data)	
0xCA	MC9S08GT60	Read A/D reply, 2 Data Bytes: ATD1RH, ATD1RL, voltage result = ((ATD1RH << 8) I ATD 1RL) * Vref / 1024	
0xCB	Serial / USB	Read VBAT reply, 2 Data Bytes: ATD1RH, ATD1RL, Refer to RFTestAp source code for data processing details	
0xCC	Serial / USB	Read Temperature & Humidity reply, Refer to RFTestAp source code for data processing details	

Figure 14.3: The reason the DLP-RF2-Z can seemingly outdo the DLP-RF1-Z is that the 20-pin female header sockets on the DLP Design sensor and relay modules can only accommodate the DLP-RF2-Z. The DLP-RF1-Z has no I/O interface other than the single USB interface.

Other than the way they interface, the DLP-RF1-Z and DLP-RF2-Z are very similar in design. However, you can obviously see that the DLP-RF1-Z's role is data collector, while the DLP-RF2-Z will most always be found at the remote end of the link.

Figure 14.4 is a look at the DLP-RF2-Z's 20-pin interface.

Once you have the framework in place, writing application code for the DLP-RF2-Z is as easy as implementing the DLP-RF2-Z hardware. A piece of code designed to run on a Microchip PIC microcontroller is presented in Code Snippet 14.1. This piece of code sets one of the latching relays on the DLP Design DLP-RF2-ZRELAY module.

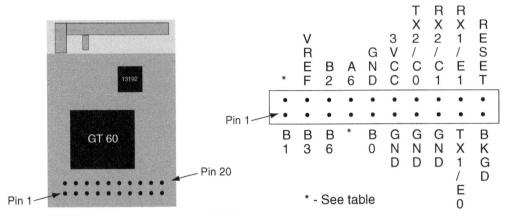

Top View (Interface Header on bottom of PCB)

Pin #	Header Pin Description
1	**PTB1** (I/O) Port Pin B1 connected to the microcontroller; A/D Channel 1
2	**DLP-RF2: PTB4** (I/O) Port Pin B4; A/D Channel 4 **DLP-RF2-Z: PTA1** (I/O) Port Pin A1
3	**PTB3** (I/O) Port Pin B3 connected to the microcontroller; A/D Channel 3
4	Vref for A/D Converter (2.08V-VCC)
5	**PTB6** (I/O) Port Pin B6 connected to the microcontroller; A/D Channel 6
6	**PTB2** (I/O) Port Pin B2 connected to the microcontroller; A/D Channel 2
7	**DLP-RF2: PTB5** (I/O) Port Pin B5; A/D Channel 5 **DLP-RF2-Z: PTA2** (I/O) Port Pin A2
8	**PTA6** (I/O) Port Pin A6 connected to the microcontroller
9	**PTB0** (I/O) Port Pin B0 connected to the microcontroller; A/D Channel 0
10,11,13,15	**Ground**
12	**Power Supply**—connect external power supply: 2.0 (MIN) to 3.4 Volts (MAX)
14	**PTC0** (I/O) Port Pin C0 connected to the microcontroller; TxD2
16	**PTC1** (I/O) Port Pin C1 connected to the microcontroller; low power enable for SIPP firmware if held low at reset/power up
17	**PTE0** (I/O) Port Pin E0 connected to the microcontroller; TxD1
18	**PTE1** (I/O) Port Pin E1 connected to the microcontroller; RxD1
19	**BKGD** Background Debug
20	**RESET#** Microcontroller Reset Input

Figure 14.4: Translate to a DLP-RF2-Z. The compact 20-pin interface makes adding the DLP-RF2-Z to a microcontroller platform a very easy task.

Code Snippet 14.1
```
*****************************************************************************
void init_USART1(void);
unsigned char CharInQueue(void);
void putch(unsigned char c);
int recvchar(void);
int sendchar(int data);
```

```
unsigned char setrelay_1 (void);

#define BAUD1     9600           //desired baud rate
#define FOSC      20000000       //oscillator frequency
#define DIVIDER1 ((unsigned int)(FOSC/(16 * BAUD1) -1))
//1,2,4,8,16,32,64,128 or 256 bytes
#define USART_RX_BUFFER_SIZE  128
#define USART_RX_BUFFER_MASK ( USART_RX_BUFFER_SIZE - 1 )
// 1,2,4,8,16,32,64,128 or 256 bytes
#define USART_TX_BUFFER_SIZE  128
#define USART_TX_BUFFER_MASK ( USART_TX_BUFFER_SIZE - 1 )

unsigned char USART_RxBuf[USART_RX_BUFFER_SIZE],USART_TxBuf[USART_TX_BUFFER_
SIZE];
unsigned char USART_TxHead,USART_TxTail,USART_RxHead,USART_RxTail;

//******************************************************************
//* Init USART Function
//******************************************************************
void init_USART1(void)
{
  SPBRG1 = DIVIDER1;        //load baud rate divisor
  TRISC7 = 1;              //receive pin
  TRISC6 = 0;              //transmit pin
  TXSTA1 = 0x04;           //high speed baud rate
  RCSTA1 = 0x80;           //enable serial port and serial port pins
  USART_RxTail = 0x00;     //flush receive buffer
  USART_RxHead = 0x00;
  USART_TxTail = 0x00;     //flush transmit buffer
  USART_TxHead = 0x00;
  RC1IP = 1;               //receive interrupt = high priority
  TX1IP = 1;               //transmit interrupt = high priority
  RC1IE = 1;               //enable receive interrupt
  PEIE = 1;                //enable all unmasked peripheral interrupts
  GIE = 1;                 //enable all unmasked interrupts
  CREN1 = 1;               //enable USART1 receiver
  TX1IE = 0;               //disable USART1 transmit interrupt
  TXEN1 = 1;               //transmitter enabled
}

void interrupt USART(void)
{
  unsigned char data,tmphead,tmptail;

  if(RC1IF)
 {
   data = RCREG1;                          // read the received data
                                           // calculate buffer index
   tmphead = ( USART_RxHead + 1 ) & USART_RX_BUFFER_MASK;
   USART_RxHead = tmphead;                 // store new index

   if ( tmphead == USART_RxTail )
```

```
    {
       // ERROR! Receive buffer overflow
    }

  USART_RxBuf[tmphead] = data;            // store received data in buffer
  }


  if(TRMT1)
  {
                                          // check if all data is transmitted
   if ( USART_TxHead != USART_TxTail )
   {
                                          // calculate buffer index
      tmptail = ( USART_TxTail + 1 ) & USART_TX_BUFFER_MASK;
      USART_TxTail = tmptail;            // store new index

      TXREG1 = USART_TxBuf[tmptail];      // start transmition
    }
    else
    {
      TX1IE = 0;                          // disable TX interrupt
    }
  }
}

int recvchar(void)
{
  unsigned char tmptail;
                                          // wait for incomming data
  while ( USART_RxHead == USART_RxTail );
                                          // calculate buffer index
  tmptail = ( USART_RxTail + 1 ) & USART_RX_BUFFER_MASK;
  USART_RxTail = tmptail;               // store new index

  return USART_RxBuf[tmptail];           // return data
}

int sendchar(int data)
{
  unsigned char tmphead;
                                          // calculate buffer index
  tmphead = ( USART_TxHead + 1 ) & USART_TX_BUFFER_MASK;
                                          //wait for free space in buffer
  while ( tmphead == USART_TxTail );
                                          // store data in buffer
  USART_TxBuf[tmphead] = (unsigned char)data;
  USART_TxHead = tmphead;               // store new index

  TX1IE = 1;                             // enable TX interrupt
```

```
    return data;
}

unsigned char CharInQueue(void)
{
  return(USART_RxHead != USART_RxTail);
}

unsigned char setrelay_1 (void)
{
 unsigned char counter, rc;
 unsigned int destaddr = 0x0003;
 unsigned int srcaddr = 0x0002;
 unsigned char tx_buffer[128];
 unsigned char rx_buffer[128];

 //init packet index pointer
 unsigned int buffer_ptr =1;
 //Destination ID MSB
 tx_buffer [buffer_ptr++] = (unsigned char)((destaddr & 0xff00) >>8);
 //Destination ID LSB
 tx_buffer [buffer_ptr++] = (unsigned char)(destaddr & 0x00ff);
 //Source ID MSB
 tx_buffer [buffer_ptr++] = (unsigned char)((srcaddr & 0xff00) >>8);
 //Source ID LSB
 tx_buffer [buffer_ptr++] = (unsigned char)(srdaddr & 0x00ff);
 //Command byte: Relay
 tx_buffer [buffer_ptr++] = 0xB8;
//Select Relay 1
  tx_buffer [buffer_ptr++] = 0x01;
 //Set Relay
 tx_buffer [buffer_ptr++] = 0x01;
 //assign number of bytes in packet to array position zero

 tx_buffer [0] = buffer_ptr-1;
//send packet in tx_buffer out of serial port
 for(counter = 0;counter<buffer_ptr;++counter)
    sendchar(tx_buffer[counter];
 while(!CharInQueue());          //wait for 0xC0 return byte
 rc = recvchar();               //get incoming character
 return(rc);
}
```

****

Code Snippet 14.1: This is a representative application of the DLP-RF2-Z. The reply to this command is analyzed by the main application to make sure the DLP-RF2-ZRELAY module returned the correct return code.

The code in Code Snippet 14.1 uses a DLP-RF2-Z to send a relay set command to another DLP-RF2-Z mounted on a DLP-RF2-ZRELAY. I just happen to have a DLP-RF2-ZRELAY in my possession and I snapped a photograph of it for you, shown in Photo 14.3.

Photo 14.3: The wiring quick connect you see in the top left quadrant is a door switch input. When a switch attached to this connector is disturbed, the DLP-RF2-Z mounted on the DLP-RF2-ZRELAY wakes up and sends a broadcast packet.

As you can see at the top of Photo 14.3, the DLP-RF2-ZRELAY module is designed to carry a DLP-RF2-Z radio module. When the DLP Design module DLP-RF2-Z/DLP-RF2-ZRELAY combination is powered up, the system goes immediately into low-power mode, drawing less than 40 microamps of current. The DLP-RF2-Z can be programmed to wake up periodically to check in with the controller or the door switch can act as a catalyst to awaken the relay system to file a report. The DLP-RF2-ZRELAY's relays are latching relays, which are controlled by pulses emanated by the Freescale Semiconductor microcontroller onboard the DLP-RF2-Z. Take another look at the SIPP commands to see what functionality can be gleaned from the DLP-RF2-ZRELAY module using simple serial commands.

The IEEE 802.15.4 network was designed for low-rate data collection and the DLP Design DLP-RF2SENS you see in Photo 14.4 fits the job description.

Photo 14.4: The actual temperature/humidity sensor lies directly below R9, which is just below the 20-pin DLP-RF2-Z header. Yep, that's a door switch connector at the bottom left.

I've seen that temperature/humidity sensor somewhere before. As a matter of fact, I have a few of them in the lab. Photo 14.5 is a spy satellite view of the Sensirion SHT15 temperature/humidity sensor. I'll tell you what I know about it.

Photo 14.5: The Sensirion SHT15 is a really neat device to play with. It's good for serious stuff, too.

The Sensirion SHT15 comes in a dual-sensor configuration, which provides a calibrated digital output. The humidity sensor component of the Sensirion SHT15 is based on a capacitive polymer sensing element. The bandgap PTAT (Proportional To Absolute Temperature) temperature sensor component and relative humidity sensor component are connected to an on-chip 14-bit analog-to-digital converter. Temperature and humidity data are transferred via the Sensirion SHT15's on-chip 2-wire serial interface. To provide maximum stability, all of the Sensirion SHT15 sensing and communications elements are deposited on a single CMOS chip. The Sensirion SHT15's accuracy is ensured by factory-programmed calibration coefficients, which are used by the Sensirion SHT15 internally to calibrate signals from the temperature and humidity sensor elements when measurements are made.

Although Photo 14.5 implies that the Sensirion SHT15 is an 8-pin device, it is actually a 4-pin device with pins for power, ground, data (DATA) and clock (SCK). The Sensirion SHT15 can operate within a supply voltage range of 2.4VDC to 5.5VDC.

The terms 2-wire, DATA and SCK are synonymous with I²C. However, the Royal Philips invention is not used by the Sensirion SHT15. The Sensirion SHT15 uses a proprietary bit-bang approach to deliver its data. The SCK signal is used to synchronize the data transfers between the SHT15 and the DLP-RF2SENS's Freescale Semiconductor microcontroller. The Sensirion SHT15 datasheet lays out the details of the communications interface. The good news is that you don't have to write a single line of Sensirion SHT15 code to use the Sensirion SHT15 in your applications, as the SIPP firmware provides full support for the Sensirion SHT15 temperature/humidity sensor.

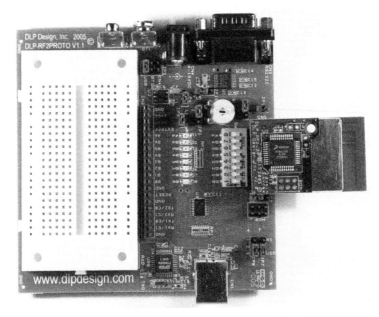

Photo 14.6: Here's an aerial view of the DLP-RF2PROTO with a DLP-RF2-Z loaded into the 20-pin female header position. The DLP-RF2-Z's interface pins are brought out to the inline female header you see directly to the right of the solderless breadboard.

As if implementing isn't easy enough, Don has come up with the DLP-RF2PROTO. The DLP-RF2PROTO is a prototyping platform for the DLP-RF2-Z. As you can see in Photo 14.6, the DLP-RF2PROTO has both a serial and USB interface, which are power and I/O switchable via jumpers, a 20-pin nest for the DLP-RF2-Z and a solderless breadboard with enough capacity to support a microcontroller in the standard 40-pin DIP package. The DLP-RF2PROTO's circuitry can be powered by 9V battery, via the USB port or by an external wall wart. All you have to do is plug in your stuff, avoid letting the smoke out and develop your DLP-RF2-Z application.

Making a small modification to the command byte sequence in Code Snippet 14.1 produced the pair of Daintree Networks SNA sniffs in Sniffer Capture 14.1.

Sniffer Capture 14.1

```
************************************************************************
Frame 3 (Length = 8 bytes)
        Time Stamp: 18:46:33.764
        Frame Length: 8 bytes
        Capture Length: 8 bytes
        Link Quality Indication: 164
IEEE 802.15.4
        Frame Control: 0x0005
                .... .... .... .101  = Frame Type: Reserved (0x0005)
                .... .... .... 0...  = Security Enabled: Disabled
                .... .... ...0 ....  = Frame Pending: No more data
                .... .... ..0. ....  = Acknowledgment Request: Acknowledgment
not required
                .... .... .0.. ....  = Intra PAN: Not within the PAN
                .... ..00 0... ....  = Reserved
                .... 00.. .... ....  = Destination Addressing Mode: PAN
identifier and address field are not present (0x0000)
                ..00 .... .... ....  = Reserved
                00.. .... .... ....  = Source Addressing Mode: PAN identifier
and address field are not present (0x0000)
        Sequence Number: 2
        Frame Check Sequence: Correct

0000:   05 00 02 00 01 a0 .. ..                          ...... ..


///////////////////////////////////////////////////////////////////////////

Frame 4 (Length = 8 bytes)
        Time Stamp: 18:46:33.770
        Frame Length: 8 bytes
        Capture Length: 8 bytes
        Link Quality Indication: 120
IEEE 802.15.4
        Frame Control: 0x0005
                .... .... .... .101  = Frame Type: Reserved (0x0005)
                .... .... .... 0...  = Security Enabled: Disabled
```

```
            .... .... ...0 ....   = Frame Pending: No more data
            .... .... ..0. ....   = Acknowledgment Request: Acknowledgment
not required
            .... .... .0.. ....   = Intra PAN: Not within the PAN
            .... ..00 0... ....   = Reserved
            .... 00.. .... ....   = Destination Addressing Mode: PAN
identifier and address field are not present (0x0000)
            ..00 .... .... ....   = Reserved
            00.. .... .... ....   = Source Addressing Mode: PAN identifier
and address field are not present (0x0000)
        Sequence Number: 1
        Frame Check Sequence: Correct
```

```
0000:   05 00 01 00 02 c0 .. ..                      .....@..
********************************************************************************
```

Sniffer Capture 14.1: This is an easy sniff sequence to solve. Just use the packet layout of Figure 14.1 and find the command in Figure 14.2.

Frame 3 of Sniffer Capture 14.1 is a five-byte message sent from a DLP-RF1-Z to a DLP-RF2-Z. How do I know that? Easy, the factory default ID for a DLP-RF2-Z is 0x02 and the default ID for a DLP-RF1-Z is 0x01. The command is a PING (0x0A).

```
********************************************************************************
0000:   05 00 02 00 01 a0 .. ..    PING from DLP-RF1-Z to DLP-RF2-Z

0000:   05 00 01 00 02 c0 .. ..    REPLY from DLP-RF2-Z to DLP-RF1-Z
********************************************************************************
```

Frame 4 of Sniffer Capture 14.1 is yet another five-byte message in the form of a reply denoted by the SIPP reply command 0xC0.

Fortunately, I've had prior experience with DLP Design's USB product line. My feeling about DLP Design is that if you really need to get a serious application up and working fast, get some DLP Design stuff to do it with. With every DLP Design product I've worked with, I've found that, once you get what you think you want, you can always take it a level further down the complexity scale if you choose to. That fact is evident here as, if you don't see what you want in the modules you and I have talked about in this chapter, DLP Design can either provide you with the technical details you require or build it to suit you.

About DLP Design

As I mentioned earlier, I've done some magazine things with Don's stuff prior to the writing of this chapter. Don Powrie's name can be found at the top of a few magazine columns as well. Take it from one who knows first-hand—if you're proud enough of a product you sell to put your name on an international magazine column about that product, it had better work and work very well. I don't think Don will have anything to worry about.

If you like bluegrass music, then you can thank Bill Monroe. Bill's Blue Grass Boys put blue-grass music into the mainstream. In fact, the bluegrass moniker is derived from Bill's band's name. If you have absolutely no idea what bluegrass music is, check out the movie *O Brother Where Art Thou*. Trust me, you will be hooked.

The smooth and jazzy Nashville sound was ushered in by the likes of Chet Atkins. Outlaw music opposed the slick Nashville-produced music. Can you name any of the "outlaws"?

Microchip

The IEEE 802.15.4 hardware for this IEEE 802.15.4 MAC/PHY discussion is marketed by the Microchip Corporation as the PICDEM Z Demonstration Kit. The Microchip ZigBee network hardware is based on their brand-new MRF24J40, which is an IEEE 802.15.4–compliant/ZigBee-ready 2.5-GHz RF transceiver. I got my hands on one and put a camera lens against it in Photo 15.1.

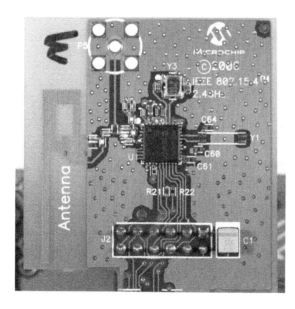

Photo 15.1: I don't know if you can make it out, but my MRF24J40 module is so new that it carries an engineering sample of the part.

The best way to show you how the MRF24J40 works is to assemble and run a simple IEEE 802.15.4 network. Our dual-node IEEE 802.15.4 network consists of a ZigBee Coordinator and a ZigBee End Device. Each of the devices is loaded with the latest version Microchip ZigBee Stack, which you can obtain freely from the Microchip web site.

The ZigBee Coordinator and ZigBee End Device in our IEEE 802.15.4 network are actually PIC18LF4620 microcontrollers, which are loaded with ZigBee Coordinator and ZigBee End Device driver code that is able to interface with the Microchip MRF24J40 radio modules that are attached to each of the ZigBee nodes. So, the network I've assembled is actually an IEEE 802.15.4-compliant network with ZigBee protocol capability. A representative Microchip ZigBee node is shown in Photo 15.2.

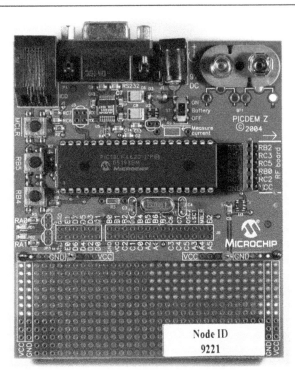

Photo 15.2: The MRF24J40 mounts in the dual-gender connector at the far right of this shot.

Let's fire everything up and see what happens.

Birth of a Microchip ZigBee Network

The ZigBee source code contains the following line of code:

```
*******************************************************************************
#define ALLOWED_CHANNELS 0x00001000
*******************************************************************************
```

I know from that line of code that the PIC-based ZigBee application I'm about to run will use channel 12, which, as you already know, lies in the channel assignments for the 2.4-GHz band.

The very first thing the Microchip ZigBee Coordinator driver firmware does is to initialize the PICDEM Z PIC microcontroller hardware with a call to the *HardwareInit* function from the *Coordinator.c* module. The PIC hardware initialization process involves setting up the PIC microcontroller's SPI (Serial Peripheral Interface) portal and the direction (input or output) and initial output states (1 or 0) of the PIC microcontroller's general-purpose I/O pins. The PICDEM Z's host PIC18LF4620 microcontroller's SPI engine must be activated and configured as the MRF24J40 IEEE 802.15.4–compliant 2.5-GHz RF transceiver communicates with the PIC microcontroller using the SPI portal. In addition, the PIC microcontroller's Master SPI interface must use an additional I/O pin to act as the SPI select line for the MRF24J40, which is acting as an SPI slave. The code snippet from the

HardwareInit function pertains to the MRF24J40 2.4-GHz transceiver is shown in Code Snippet 15.1.

Code Snippet 15.1

```
********************************************************************************
void HardwareInit(void)
{


//------------------------------------------------------------------------
    // This section is required to initialize the PICDEM Z for the MRF24J40
    // and the ZigBee Stack.

//------------------------------------------------------------------------

    SPIInit(); // a definition that simply sets SSP1IF  = 1;

    #if (RF_CHIP == MRF24J40)
        // Start with MRF24J40 disabled and not selected
        PHY_CS              = 1;
        PHY_RESETn          = 1;

        // Set the directioning for the MRF24J40 pin connections.
        PHY_CS_TRIS         = 0;
        PHY_RESETn_TRIS     = 0;

        // Initialize the interrupt.
        INTCON2bits.INTEDG0 = 0;
    #endif

    // Initialize the SPI pins and directions
    LATC3               = 1;    // SCK
    LATC5               = 1;    // SDO
    TRISC3              = 0;    // SCK
    TRISC4              = 1;    // SDI
    TRISC5              = 0;    // SDO

    // Initialize the SPI module
    SSPSTAT = 0xC0;
    SSPCON1 = 0x20;

//------------------------------------------------------------------------
    // This section is required for application-specific hardware
    // initialization.

//------------------------------------------------------------------------

    // D1 and D2 are on RA0 and RA1 respectively, and CS of the TC77 is on
RA2.
    // Make PORTA digital I/O.
    ADCON1 = 0x0F;
```

```
    // Deselect the TC77 temperature sensor (RA2)
    LATA = 0x04;

    // Make RA0, RA1, RA2 and RA4 outputs.
    TRISA = 0xE0;

    // Clear the RBIF flag (INTCONbits.RBIF)
    RBIF = 0;

    // Enable PORTB pull-ups (INTCON2bits.RBPU)
    RBPU = 0;

    // Make the PORTB switch connections inputs.
    TRISB4 = 1;
    TRISB5 = 1;
}
```
****

Code Snippet 15.1: If you're not familiar with the innards of a PIC, get the PIC18LF4620 datasheet from Microchip and you'll be able to match up all of the mnemonics in this Code Snippet. I'll get you started. LATX are output pins and TRISXX sets the general-purpose I/O port direction. A "1" makes the port pin an input and a "0" puts the port pin in output mode.

Once all of the PIC18LF4620's general-purpose I/O pins and SPI stuff are pointing in the right direction, the *ZigBeeInit* function, which is located in the *ZigBeeTasks.c* module, is invoked from within the *Coordinator.c* code module. As you can see in Code Snippet 15.2, all of the normal initialization tasks you come to expect by now (if you haven't been chapter skipping) come into play.

Code Snippet 15.2
****

```
 void ZigBeeInit(void)
{
    SRAMInitHeap();

    MACInit();
    NWKInit();
    APSInit();
    ZDOInit();

    TxHeader = 127;
    TxData = 0;
    RxWrite = 0;
    RxRead = 0;

    // Set up the interrupt to read in a data packet.
    // set to capture on falling edge
#if (RF_CHIP == UZ2400) || (RF_CHIP == MRF24J40)
    CCP2CON = 0b00000100;
```

```
#elif (RF_CHIP == CC2420)
    CCP2CON = 0b00000101;
#endif

    // Set up the interrupt to read in a data packet.
#if (RF_CHIP==UZ2400) || (RF_CHIP == MRF24J40)
    INT0IF = 0;
    INT0IE = 1;
#elif (RF_CHIP==CC2420)
    CCP2IF = 0;
    CCP2IP = 1;
    CCP2IE = 1;
#endif
    InitSymbolTimer();

    ZigBeeStatus.nextZigBeeState = NO_PRIMITIVE;
    CurrentRxPacket = NULL;
}
```
****

Code Snippet 15.2: Note the inclusion of other types of IEEE 802.15.4 radio types. You should be familiar with one of them.

At this point, the Coordinator is PAN-less. The Microchip ZigBee stack works with states and NO_PRIMITIVE is the current state, which was set in the initialization routine shown in Code Snippet 15.2. This will lead to the application code falling into the set of statements shown in Code Snippet 15.3.

Code Snippet 15.3
****

```
if (!ZigBeeStatus.flags.bits.bTryingToFormNetwork)
{
  params.NLME_NETWORK_FORMATION_request.ScanDuration        = 8;
  params.NLME_NETWORK_FORMATION_request.ScanChannels.Val     = ALLOWED_
CHANNELS;
  params.NLME_NETWORK_FORMATION_request.PANId.Val            = 0xFFFF;
  params.NLME_NETWORK_FORMATION_request.BeaconOrder          = MAC_PIB_
macBeaconOrder;
  params.NLME_NETWORK_FORMATION_request.SuperframeOrder      = MAC_PIB_
macSuperframeOrder;
  params.NLME_NETWORK_FORMATION_request.BatteryLifeExtension = MAC_PIB_
macBattLifeExt;
  currentPrimitive = NLME_NETWORK_FORMATION_request;
}
```
****

Code Snippet 15.3: Don't worry, you and I both know we'll find out what the MAC PIB values are when we examine the Daintree Networks SNA sniff. Note that we're passing a request primitive here.

The ZigBeeStatus.flags.bits.bTryingToFormNetwork bit is set to indicate that a PAN is in the attempt stage and the following code is executed:

```
*******************************************************************
params.MLME_SCAN_request.ScanType = MAC_SCAN_ENERGY_DETECT;
// ScanChannels is already in place
// ScanDuration is already in place
return MLME_SCAN_request;
*******************************************************************
```

The commented ScanChannels and ScanDuration lines are reminders that the values for those attributes were passed with the primitive call. We know that only a single channel (12) will be scanned. Otherwise, the Microchip ZigBee stack would go through iterations to determine the best channel on which to establish the PAN. As you already know, things coming down from the NWK layer pass through the appropriate SAP to the MAC layer, which is what we see happening in Code Snippet 15.4.

Code Snippet 15.4

```
*******************************************************************
params.MLME_START_request.LogicalChannel     = nwkStatus.discoveryInfo.
channelList[nwkStatus.discoveryInfo.currentIndex].channel;
params.MLME_START_request.BeaconOrder        = MAC_PIB_macBeaconOrder;
params.MLME_START_request.SuperframeOrder    = MAC_PIB_macSuperframeOrder;
params.MLME_START_request.fields.Val         = MLME_START_IS_PAN_COORDINATOR;
params.MLME_START_request.fields.bits.BatteryLifeExtension = macPIB.
macBattLifeExtPeriods;
return MLME_START_request;
*******************************************************************
```

Code Snippet 15.4: At this point, we've scanned and selected a channel on which to establish the PAN. Note the passage of the same MAC PIB attribute values from the NWK NLME to the MAC MLME.

A good thing to have happen at this point is to process the MLME_START_confirm primitive, which, if the confirm message indicates a successful start, will kick off the NLME_NETWORK_FORMATION_confirm operation. A PAN exists.

Someone has to join the PAN for anything useful to happen. You all know what happens next:

```
*******************************************************************
params.NLME_PERMIT_JOINING_request.PermitDuration = 0xFF;    // No Timeout
currentPrimitive = NLME_PERMIT_JOINING_request;
*******************************************************************
```

The NWK, MAC and PHY layers have established a PAN at the behest of the application. The response to the Beacon Request sniff in Sniffer Capture 15.1 should be easily absorbed.

Sniffer Capture 15.1

```
*******************************************************************
Frame 3 (Length = 16 bytes)
       Time Stamp: 19:22:04.298
       Frame Length: 16 bytes
       Capture Length: 16 bytes
       Link Quality Indication: 112
```

```
IEEE 802.15.4
      Frame Control: 0x8000
                  .... .... .... .000  = Frame Type: Beacon (0x0000)
                  .... .... .... 0...  = Security Enabled: Disabled
                  .... .... ...0 ....  = Frame Pending: No more data
                  .... .... ..0. ....  = Acknowledgment Request: Acknowledgment
not required
                  .... .... .0.. ....  = Intra PAN: Not within the PAN
                  .... ..00 0... ....  = Reserved
                  .... 00.. .... ....  = Destination Addressing Mode: PAN
identifier and address field are not present (0x0000)
                  ..00 .... .... ....  = Reserved
                  10.. .... .... ....  = Source Addressing Mode: Address field
contains a 16-bit short address (0x0002)
      Sequence Number: 1
      Source PAN Identifier: 0x3f80
      Source Address: 0x0000
      MAC Payload
            Superframe Specification: 0xcfff
                      .... .... .... 1111  = Beacon Order (0x000f)
                      .... .... 1111 ....  = Superframe Order (0x000f)
                      .... 1111 .... ....  = Final CAP Slot (0x000f)
                      ...0 .... .... ....  = Battery Life Extension: Disabled
                      ..0. .... .... ....  = Reserved
                      .1.. .... .... ....  = PAN Coordinator: Transmitter is
a PAN Coordinator
                      1... .... .... ....  = Association Permit: Coordinator
accepting Association Requests
            GTS Specification: 0x00
                      .... .000  = GTS Descriptor Count (0x00)
                      .000 0...  = Reserved
                      0... ....  = GTS Permit: Coordinator not accepting GTS
Requests
            Pending Address Specification: 0x00
                      .... .000  = Number of short Addresses pending: 0
                      .... 0...  = Reserved
                      .000 ....  = Number of extended Addresses pending: 0
                      0... ....  = Reserved
            Beacon Payload
                  Protocol ID: ZigBee NWK (0x00)
      Frame Check Sequence: Correct
NWK Layer Information: 0x8411
            .... .... .... 0001  = Stack Profile (0x1)
            .... .... 0001 ....  = nwkcProtocolVersion (0x1)
            .... ..00 .... ....  = Reserved (0x0)
            .... .1.. .... ....  = Router Capacity: True
            .000 0... .... ....  = Device Depth (0x0)
            1... .... .... ....  = End Device Capacity: True
```
****

Sniffer Capture 15.1: It's almost like a Western—the cowboy arrives to save the day and then rides off into the sunset. In this case the ZigBee End Device was accepted for association and you should know the rest of the story.

If you're not proficient in the ways of IEEE 802.15.4 and ZigBee network flows by now, please hand this book to your sister. On the other hand, if you feel good about your IEEE 802.15.4 and ZigBee network flow expertise, let's get off the beaten path and look at something else Microchip has to offer that smells like ZigBee. If you stiffed through my first two books and you've made it this far through this one, you know that when it comes to doing things Ethernet, I relied heavily in both cases on my Network General Sniffer Portable application suite. I used the Sniffer in the development of all of my wired and wireless microcontroller Ethernet designs. In fact, I used the Sniffer captures to lay open the 802.11b and 802.3 specifications bit by bit in the texts. The Network General Sniffer Portable tool in its many variants works quite well for Ethernet development.

I don't know what kind of thoughts you may have had about the ZigBee protocol before you started reading this book, as on the outside, ZigBee appears to be a simple wireless implementation that should be easily grasped and applied. However, as you have seen in our discussions, on the inside, ZigBee can be quite complex. That's why I used Daintree Networks SNA captures so heavily to convey the basic concepts of IEEE 802.15.4 and ZigBee networking. Now that you have a good understanding of the flow of things in IEEE 802.15.4 and ZigBee networks, you can learn a great deal about the network layer and primitives by following the Microchip ZigBee stack thread I described previously.

The free Microchip ZigBee stack and the MRF24J40 are not the only things Microchip that are ZigBee related. If you choose to develop your IEEE 802.15.4 or ZigBee application with Microchip products, Microchip makes another ZigBee development tool available to you called ZENA. Let's fire up the application we just discussed again and look at it through the eyes of ZENA.

ZENA

Nope…ZENA isn't one of those scantily clad female superheroes that can cut bad guys into pieces of sushi with her good looks, quick hands and sharp blade. ZENA, formally known as the ZENA Wireless Network Analyzer, is actually a piece of hardware and some supporting software that performs a similar function to the Daintree Networks SNA application you've seen all throughout this book.

ZENA's network support provides for decoding ZigBee packets in the 2.4-GHz frequency spectrum. As ZigBee networks are formed, ZENA analyzes the network activity and draws the network topology as it is created. ZigBee packet transactions can be viewed in real time and can be recorded for variable speed playback. In addition to ZigBee 802.15.4 packet sniffing/packet analyzer duty, ZENA contains a configuration tool that can be used to act on the free Microchip ZigBee stack to help ZigBee developers modify the Microchip ZigBee stack to suit their application.

The ZENA Wireless Network Analyzer hardware you see in Photo 15-3 is based on a PIC18LF2550, which is the little brother of the USB-equipped PIC18LF4550. ZENA's USB interface allows the ZENA hardware to be powered directly from the USB bus via a USB mini-B cable that comes with the ZENA Wireless Network Analyzer package. Information flows between the ZENA and host personal computer within the HID standard class.

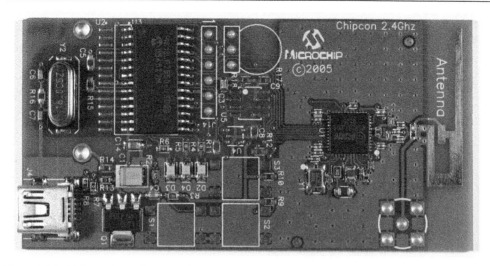

Photo 15.3: This version of ZENA is built around the familiar CC2420 from Texas Instruments/ Chipcon. The new version of ZENA will revolve around the new Microchip MRF24J40 2.4-GHz transceiver IC.

Summoning ZENA

The Microchip ZigBee stack comes with a couple of ready-to-roll demo projects. I want to show you how the ZENA Wireless Network Analyzer works and I really don't need a fancy ZigBee application to do that. So, I used Microchip's C18 C compiler to build up the basic demo ZigBee Coordinator and ZigBee RFD (Reduced Function Device) projects.

Just to make sure that everything was really working before I invoked ZENA, I attached the ZigBee Coordinator node to an instance of my preferred terminal emulator, Tera Term Pro. I powered up the ZigBee Coordinator node and received a positive response. The newly loaded ZigBee Coordinator node did indeed perform a channel scan and form a new PAN. At this point, my confidence level was high enough to go ahead and build the code for and load the ZigBee RFD node. Just for grins I fired up the ZigBee RFD node and it joined the existing PAN that had just been created by the ZigBee Coordinator node. The Microchip PICDEM Z ZigBee hardware is ready to go.

With known-good ZigBee nodes on the ready, I loaded the ZENA Wireless Network Analyzer code into my laptop. After starting the ZENA application, I selected Network Monitor mode and restarted the ZigBee Coordinator node. You and I already know that the ZigBee nodes are operating on channel 12 and I set ZENA up accordingly and refired the ZigBee PAN Coordinator node. At this point, the ZigBee Coordinator has searched for other Zig-Bee Coordinators that may be operating in the same channel range. Obviously, in our case there are no other ZigBee Coordinators. Thus, the lone ZigBee Coordinator establishes a network and assigns it a unique 16-bit PAN ID. I've combined all of the good stuff from the ZigBee Protocol window and the ZigBee Network Monitor console window into a view in Screen Capture 15.1.

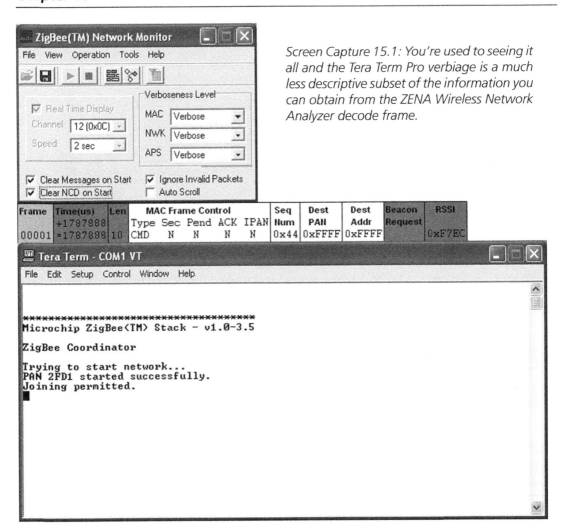

Screen Capture 15.1: You're used to seeing it all and the Tera Term Pro verbiage is a much less descriptive subset of the information you can obtain from the ZENA Wireless Network Analyzer decode frame.

Note that the new PAN ID is revealed in the Tera Term Pro window. The PAN ID revelation in the Tera Term Pro window is part of the ZigBee Coordinator application code. A ZigBee network is no good if there aren't any nodes to perform useful work. So, the next logical step is to enable the ZigBee RFD node I just built and see what the ZENA Wireless Network Analyzer has to show me. I've also brought up another instance of Tera Term Pro on COM2 to allow us to see what the ZigBee RFD node has to say as it is coming up.

You already have a very good idea about how devices join ZigBee networks. So, let's work through what we see in the ZENA packet decodes and the Tera Term Pro output. Once I powered up the ZigBee RFD node, as expected the node immediately tried to find an established ZigBee network. In Screen Capture 15.2, this is shown by the Beacon Request frame with a sequence number of 0x00 (Frame 00002), which was transmitted by the lonely ZigBee RFD node.

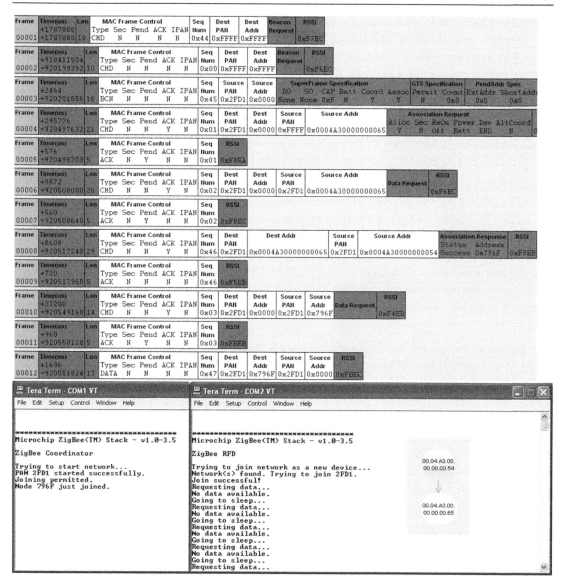

Screen Capture 15.2: The nodes that were drawn by ZENA are identified by both color and their MAC addresses. The connecting arrowed line between the ZigBee Coordinator and ZigBee RFD nodes represents the communications route between the nodes. The packet decode frames extend horizontally in the Protocol window to accept the maximum ZigBee MAC packet size of 127 bytes.

Just before the ZigBee RFD node punched a hole in the ether looking for a friend, the ZigBee Coordinator generated a Beacon Request Frame (the active channel scan indication) sequence numbered as 0x44 and represented as Frame 00001 in Screen Capture 15.2. The sequence numbers will allow us to keep up with who's talking and who's not, as the next logical Zig-Bee RFD node message will be sequence number 0x01 and the next packet from the ZigBee Coordinator should be 0x45.

In Frame 00003, the ZigBee Coordinator reveals itself by generating an informational Beacon in response to the Beacon Request issued by the wanna-be-on-my-PAN ZigBee RFD node. The Tera Term Pro COM1 output (ZigBee Coordinator) in Screen Capture 15.1 told us that the newly created PAN ID is 0x2FD1. The Tera Term Pro COM2 output (ZigBee RFD) in Screen Capture 15.2 confirms the PAN ID of 2FD1, as the ZigBee RFD node has discovered the 0x2FD1 network and is trying to join it. The PAN ID was passed in the Source PAN field of the packet and the ZigBee Coordinator's address was passed in the Source Address field. This packet is labeled with a sequence number of 0x45 (Frame 00003), which tells us that the ZigBee Coordinator is the originator of the packet.

The ZigBee RFD node issues an Association Request in Frame 00004. How do we know this? The ZENA packet decode tells us that this is an Association Request. The next clue is the sequence number, which is 0x01. Recall that the sequence numbers for the ZigBee Coordinator began with 0x44. A look at the Destination PAN field shows a value of 0x2FD1, which says the ZigBee RFD node is targeting the PAN our ZigBee Coordinator created. The ZigBee RFD node also gives up physical information about itself, including its MAC (hardware) address in this frame. The ZigBee RFD node needs to hear back from the prospective PAN Coordinator and asks for an acknowledgment in the MAC Frame Control field.

The acknowledgment for the ZigBee RFD node's Association Request (Frame 00005) is received and the ZigBee RFD node, now confident that it has found a network to join, asks for an address on PAN 0x2FD1 in Frame 00006. Following the acknowledgment in Frame 00007, the ZigBee Coordinator produces Frame 00008, which is an Association Response packet that is identified with sequence number 0x46. In this packet, the ZigBee Coordinator provides the ZigBee RFD node with the coordinator's MAC address as well as a newly assigned address for the ZigBee RFD node (0x796F). The Tera Term Pro COM2 output signals this event with a "Join successful!" message. The Tera Term Pro COM1 output celebrates this sequence of events with a "Node 796F just joined" message.

Following the ZigBee RFD node's acknowledgment of the Association Response packet in Frame 00009, the ZigBee RFD node has all of the information it needs to get to the ZigBee Coordinator and ask for data. The ZigBee RFD node repeatedly runs through a sleep and request data cycle. This sleep/request data cycle is represented by Frames 00010, 00011 and 00012 of the ZENA Screen Capture 15.2.

In the meantime, ZENA has drawn our little network up in the Network Configuration Display window. I've superimposed the actual node drawing into the Tera Term Pro COM2 window in Screen Capture 15.2.

The logic of a ZigBee data exchange and the node association process becomes evident by walking through the decoded packets as we have just done. However, there is more information to be gleaned from the packet decodes. Packet to packet elapsed time in microseconds is displayed as is the length of the packet. The MAC Frame Control fields are parsed to show such things as the type of packet (CMD, DATA, ACK, etc.) and packet status bits.

Thus far, with the help of ZENA, we've strolled through a sequence of ZigBee moves that led to the formation of a tiny ZigBee network and the association of a ZigBee RFD node. You and I should be expert at reading ZigBee sniffs by now. So, let's see if we can take a couple of ZENA-captured frames and get the gist of a ZigBee network and the data it is carrying without resorting to the inspection of the application source code. I loaded up my ZigBee nodes with a temperature application. The idea is to push a button on the ZigBee Coordinator node and get a temperature from the associated ZigBee RFD node.

Screen Capture 15.3 shows that the ZigBee RFD sent a temperature back to the ZigBee Coordinator's COM1 Tera Term Pro window. So, we can assume that all of the aforementioned network creation/network association stuff you've been reading about in the previous chapters of this book actually happened.

Frame	Time(us)	Len	MAC Frame Control					Seq Num	Dest PAN	Dest Addr	Source PAN	Source Addr
	+3888		Type	Sec	Pend	ACK	IPAN					
00024	=26270240	32	DATA	N	N	Y	N	0xFD	0x0492	0x796F	0x0492	0x0000

NWK Frame Control				Dest Addr	Source Addr	Radius	Seq Num
Type	Ver	Route	Sec				
DAT	0x1	EN	N	0x796F	0x0000	0x0A	0xB5

APS Frame Control					Dest EP	Cluster ID	Profile ID	Source EP	APS Payload	RSSI
Type	Deliv	Mode	Sec	ACK						
DAT	UNI	N/A	N	N	0x03	0x20	0xFFFF	0x09	0x11 0x00 0xE4 0x00 0x00	0xF7EC

Frame	Time(us)	Len	MAC Frame Control					Seq Num	Dest PAN	Dest Addr	Source PAN	Source Addr
	+15840		Type	Sec	Pend	ACK	IPAN					
00033	=26377872	42	DATA	N	N	Y	N	0x0B	0x0492	0x0000	0x0492	0x796F

NWK Frame Control				Dest Addr	Source Addr	Radius	Seq Num
Type	Ver	Route	Sec				
DAT	0x1	EN	N	0x0000	0x796F	0x0A	0x21

APS Frame Control					Dest EP	Cluster ID	Profile ID	Source EP	APS Payload	RSSI
Type	Deliv	Mode	Sec	ACK						
DAT	UNI	N/A	N	N	0x09	0x20	0xFFFF	0x03	0x11 0x00 0xE8 0x00 0x00 0x09 0x32 0x35 0x2E 0x38 0x37 0x35 0x30 0x20 0x43	0xEFEB

```
*******************************************
Microchip ZigBee(TM) Stack - v1.0-3.5

ZigBee Coordinator

Trying to start network...
PAN 0492 started successfully.
Joining permitted.
Node 796F just joined.
 Requesting temperature...
 Requesting temperature...
 Message sent successfully.
 Message sent successfully.
Received 25.8750 C
Received 25.8750 C
```

Screen Capture 15.3: So far, so good. This is how ZENA displays the layers. Can you pick out the MAC, NWK and APS layers in the ZENA trace?

Our known parameters consist of the received temperature (25.8750 C), the PAN ID (0492) shown in the Screen Capture 15.3 Tera Term Pro output and the node (796F) mentioned in the Screen Capture 15.3 Tera Term Pro output. I fat-fingered the switch and got a pair of identical readings. So, we'll only hack through one switch-closure packet and one tempera-ture-reading packet.

In addition to a text message from a Tera Term Pro session, what you see in Screen Capture 15.3 are a couple of typical ZigBee messages. Each ZigBee message consists of a MAC header, a NWK (Network) header, an APS (Application Support) header and the APS Payload (data). Frame 00024 is the temperature request frame and Frame 00033 contains the tempera-ture data. You've now ventured one more layer up in the ZigBee stack. Congratulations!

The PAN ID given to us in the Tera Term Pro message is confirmed by the Source and Destination PAN fields in both of the packets' MAC headers. Looking further into the MAC header we find that the ZigBee Coordinator's Source Address is 0x0000 and the ZigBee RFD node's address (Destination Address field) is 0x796F. The packet length is 32 bytes including the checksum (RSSI) and the packet has requested an acknowledgment (MAC Frame Control bit ACK set to Y).

The NWK header contains actual source and final destination of the packet. Since this is a minimal ZigBee network, the actual and finals will match up with what we've already ascer-tained from the MAC layer information.

Destination endpoint, profile and cluster information is provided by the APS header. Endpoint information seen here should correspond to endpoint configuration data within the application itself, as endpoints are directly related to application objects. From the looks of things here, the ZigBee Coordinator's endpoint 0x09 is sending data to the ZigBee RFD node's endpoint 0x03. Without examining the ZigBee RFD node's source code, we must conclude that the bytes within the APS Payload kick off the temperature measurement application within the ZigBee RFD node.

The way ZigBee jockeys data back and forth between nodes is very similar to the way IP does it. You can cross verify the source and destination addresses of the ZigBee Coordinator and ZigBee RFD nodes in Frame 00033. The same can be said for the NWK header of Frame 00033. Moving to the APS header, the endpoints also cross verify with the source and desti-nation endpoints pointing to each other.

What we really care about are the bytes inside the APS Payload area of the decode. I elimi-nated the bytes that didn't match up with an ASCII number or letter and determined that the 0x09 is the number of characters in the temperature data returned by the ZigBee RFD node. If you're not used to doing ASCII in your head, get out your ASCII chart and you will decipher "25.8750 C" as the last nine characters of the payload data.

You and I have reverse-engineered the configuration data out of the ZigBee nodes in our little network by using ZENA captures and simple logical deduction. ZENA also works the other way around. The ZENA configuration tool allows you to graphically enter the con-figuration data we saw in the ZENA capture fields. The ZigBee Device portion of the ZENA configuration tool allows the ZigBee programmer to enter device-specific information such as the node MAC address, ZigBee device type (ZigBee Coordinator, ZigBee Router or ZigBee End Device) and IEEE device type (Full Function Device/FFD or Reduced Function Device/RFD). Endpoint data can be specified in another area of the ZENA configuration tool. Once all of the configuration data has been entered, the ZENA configuration tool generates a basic definitions file for stack configuration, a ROM initialization file that includes ZigBee device descriptors and a project linker file for the PIC microcontroller you specified in the ZENA configuration tool.

About Microchip

Can you say "PIC microcontroller"? If your sticky little fingers are on the pages of this book, I'll bet that you have used a PIC microcontroller in at least one of your projects. If you don't know what a PIC is, give this book to your sister. She knows what a PIC is. Microchip is a very large corporation. However, how many corporate CEOs do you know that attend the technical conferences and actually talk to the attendees? Microchip's Steve Sanghi does. And guess what—Eric Lawson and the Microchip staff answer phone calls and respond to emails. My thanks to Eric and the entire ZigBee development team at Microchip for their contributions to the content of this book.

Willie Nelson is still an outlaw and the late Johnny Paycheck is without a doubt an outlaw, wherever his soul may be. You've probably heard the term "outlaw" applied to Waylon Jennings and Merle Haggard as well. Do you know which unmentioned outlaw loved to perform in prisons?

Telegesis

Thanks to Danny Lemos at Lemos International, the folks that supply the Telegesis ETRX2 ZigBee modules, I sat in wonderment for at least an hour just watching the mesh networking do its thing in the Daintree Networks SNA Visual Device Tree window, as I removed and re-inserted Telegesis ETRX2 nodes in the mesh network the modules had created. Light up some incense and enjoy the show, beginning with Screen Capture 16.1.

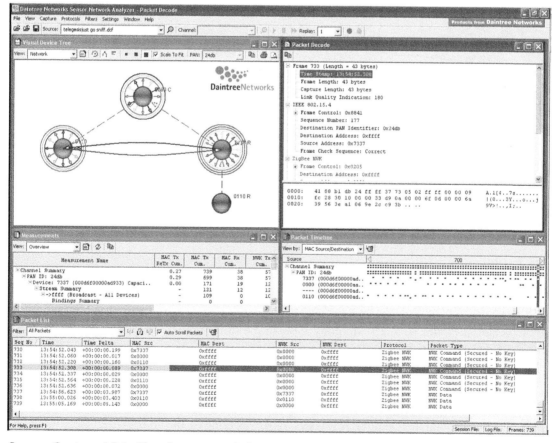

Screen Capture 16.1: The rings around the Routers and Coordinator are levels of attempted interconnectivity (broadcast levels) that each node has attained. In other words, these guys are searching the airwaves for each other and new life forms in a big way.

As you can see at the bottom of Screen Capture 16.1, the ETRX2 network is rather busy for a low-power, low-data-rate IEEE 802.15.4-based ZigBee network. The ETRX2 nodes are running at their out-of-the-box factory defaults. We can administer some Ritalin via the ETRX2's S-Registers to calm things down.

Figure 16.1 is a better view of the Daintree Networks SNA Visual Device Tree window shown in Photo 16.1. The view in Figure 16.1 is called a Tree layout and is a graphical subset of the Network View.

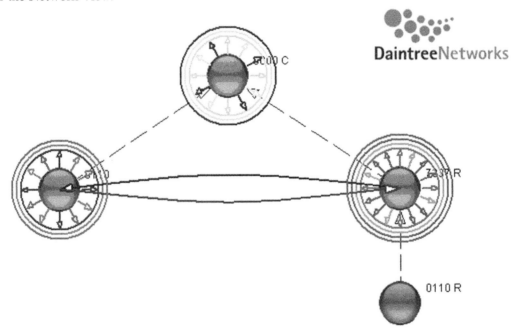

Figure 16.1: Tree layout. I powered off the Router labeled 0110R in midstream. You probably can't make out the legend, but the Router at the far left is the "recovered" version of 0110R.

The constellation you see in Figure 16.2 is still a Network View but is shown in the Radial format.

In the visual in Figure 16.3 is a Topology View of the same Telegesis network you see in Figures 16.1 and 16.2 but shown here in Tree format.

The Daintree Networks SNA Visual sniffs are really helpful when it comes to understanding what's going on in a ZigBee network. The real point I want to impress upon you here is that I did nothing other than appoint a PAN Coordinator and the rest of the work that generated the traffic I captured with the Daintree Networks SNA was done by the Telegesis ETRX2 modules.

The heart of a Telegesis ETRX2-based network is the ETRX2 ZigBee transceiver module. The Telegesis ETRX2 module is a low-power 2.4-GHz ISM band transceiver based on the single chip Ember EM250. That takes care of the hardware.

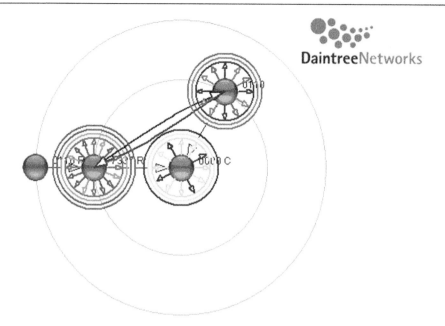

Figure 16.2: Radial format. Note the PAN Coordinator is the center of the universe and the Routers orbit in the nearest radius. ZigBee devices associated with the Routers would orbit in the ring where the dead node resides.

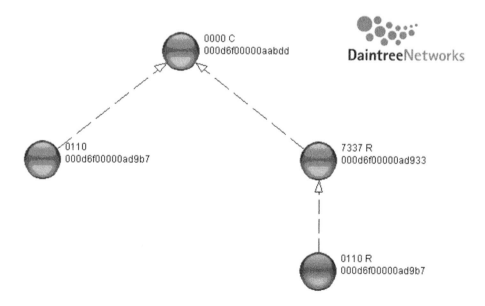

Figure 16.3: The dotted lines are association lines with the arrow pointing towards the parent. Notice that 0110R was originally associated with 7337R until I pulled the plug. I repowered 0110 much later and it reassociated with the Coordinator.

The other side of the story is the embedded EmberZNet meshing stack. The integrated Ember EM250 hardware is the reason for not having to be one of those moon-and-star-pointy-hat RF guys; the EmberZNet meshing stack eliminates having to also have a software engineering degree to go with the pointy hat. I can't show you the EmberZnet stack, but I can show you a naked ETRX2 module. Make the kids leave the room and take a peek at the disrobed ETRX2 I caught on camera in Photo 16.1.

Photo 16.1: ETRX2 module, disrobed. Very unassuming. Very powerful. Very easy to use. Very nice.

Look closely at the Telegesis logo in Photo 16.1. That's somebody sitting under a tree! If you really use your imagination, that person is also reading something. He or she is probably pouring over the Telegesis tech manuals. Hmmmm…That could be me sitting there.

Enough of the cloud watching. Let's flip this thing over. As you can see in Photo 16.2, the ETRX2's solder-side view, the ETRX2 is designed to integrate easily into most any hardware design. That's because it's really an IC disguised as a module. Just design in some ETRX2-complementary pads and you can incorporate instant ZigBee mesh networking into your project.

You don't have to build up a motherboard to develop your application with the ETRX2 unless you don't have anything else to do. Telegesis offers an ETRX2 development kit that puts the 1.27-mm header on the ETRX2's solder side, as shown in Photo 16.3.

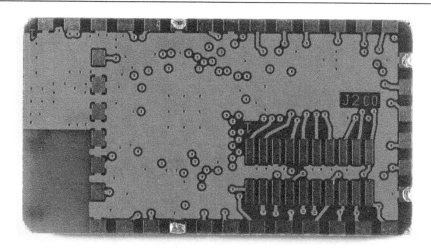

Photo 16.2: The 20-pin DIP outline can be populated with a 1.27-mm header strip, which makes the ETRX2 pluggable. Otherwise, the pads around the perimeter are the soldering points. ZigBee integration doesn't get much easier than this.

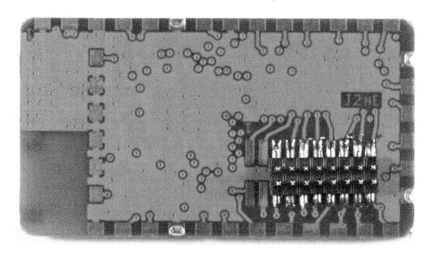

Photo 16.3: This is how the ETRX2 becomes "portable." This isn't practical for real-world applications but it does come in handy if you think you've let the smoke out of the ETRX2. Note that all of the pins are not used.

We can take this no-solder-no-build-from-scratch idea one step further and mount the ETRX2 on the Telegesis ETRX2 development kit motherboard, as I've done in Photo 16.4.

OK…That takes care of one of the ETRX2 nodes. Telegesis has made it just as easy to get running quickly on the remote end of the link as well. In Photo 16.5, the ETRX2 is mounted on a Telegesis MCB (Module Carrier Board), which is also part of the ETRX2 development kit.

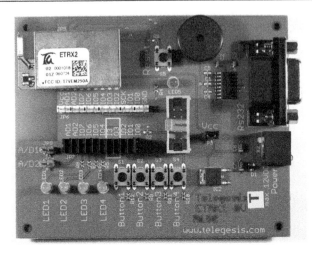

Photo 16.4: As you would expect, all of the ETRX2's general-purpose I/O lines are opened up to a header. The buttons and LEDs can be put into or taken out of the ETRX2's circuit with jumpers. There's an Atmel microcontroller in this mix somewhere, as the 10-pin header near the center of the board mates to an Atmel programming device.

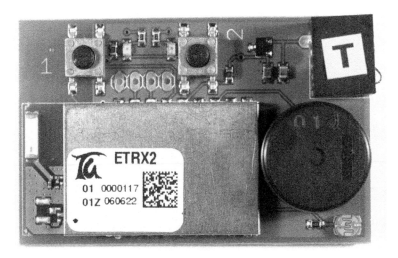

Photo 16.5: Two buttons, two LEDs, a piezo speaker and a light sensor—that's the mix on the MCB. Power for the MCB can be taken from the 4-pin header pins, which also pin out the TX and RX serial connections, or the power connector with the "T" label. The 1.27-mm connection pads are not used and the ETRX2 is mounted directly to the MCB by its solder pads.

The ETRX2 nodes can be exercised using the Telegesis Terminal program, which is a free download from the Telegesis web site. As you can see in Screen Capture 16.2, every aspect of an IEEE 802.15.4 network and flecks of ZigBee network functionality can be easily controlled by the click of a mouse button.

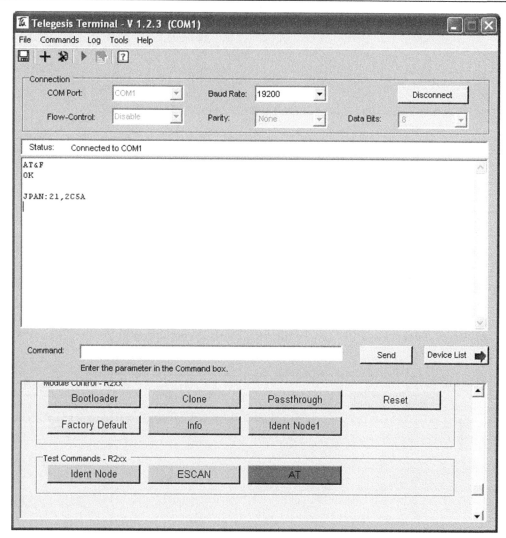

Screen Capture 16.2: I issued a Factory Default command (AT&F) and the ETRX2 mounted on the ETRX2 development board immediately jumped on starting a PAN with a PAN ID of 0x2C5A on channel 21.

Let's take control and use the Telegesis Terminal application to construct a network of our own. Instead of showing you screen shots of the Telegesis Terminal transactions, I'll translate the Telegesis Terminal screen contents to plain text. Each AT command must be followed by a carriage return/line feed sequence, which I will not show in the text translations. Be aware that the CR/LF sequences are present even though I'm not listing them for you.

We'll begin by issuing the Factory Default command once again via a click on the Factory Default button in Screen Capture 16.3.

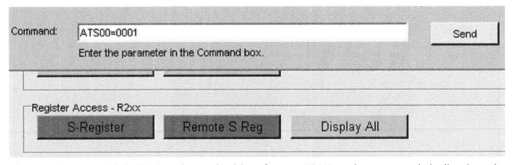

Screen Capture 16.3: The Factory Default button issues AT&F to the ETRX2.

The ETRX2 immediately attempts to start a PAN. However, I want to control the start of a PAN. So, I issue the Dissassociate Local command:

```
AT+DASSL
```

The ETRX2 responds with the message "LeftPAN". At this point, rather than allow the ETRX2 to choose a channel on which to establish the PAN, I will limit the establishment of the PAN to channel 11. This is easily done via S-Register S00, the Channel Mask register. S-Register S00 defaults to 0xFFFF, which includes all 16 of the 2.4-GHz ISM band IEEE 802.15.4 channels. If we only want to use channel 11, we must set S-Register S00 to 0x0001. Clicking on the S-Register button only partially filled in the Command window with "ATS". I had to manually enter "00=0001" before clicking on the Send button in Screen Capture 16.4.

Screen Capture 16.4: We're always looking for an "OK" to be returned, indicating the successful completion of the command.

I received an "OK" and to make sure that S-Register S00 is actually filled with 0001, I clicked on the Display All button you see in Screen Capture 16.4 and obtained the following result:

```
AT+TOKDUMP
S00:0001
S01:FFFF
```

I'm only showing S-Register S00 and S-Register S01 here as the contents of all 52 of the ETRX2's S-Registers were displayed. Obviously, AT+TOKDUMP is the AT command to dump the S-Register bank's contents. It's important to note that I received an "OK" at the end of the last S-Register value. If you're talking to a microcontroller, your firmware will have to be ready to pick up that positive acknowledgment after every successful AT command invocation. If you really don't want to deal with the "OK" prompt handing, you can disable it by clearing bit 8 of S-Register S08. In fact, you can disable a bunch of prompts in S-Registers S07 and S08.

I showed you the default setting of S-Register S01 (0xFFFF) as we're about to make a change to that value as well. S-Register S01 holds the preferred PAN ID. Let's get Brylcreme and set the PAN ID to 0x1DAB (will do ya). For the youngsters in the audience, Brylcreme was a men's hair cream and their commercial jingle was centered around "one dab will do ya." Anyway, I executed the contents of the S-Register command shown in Screen Capture 16.5.

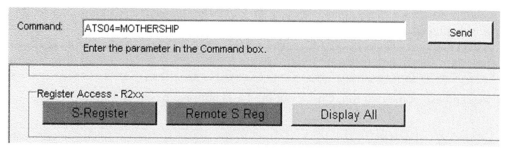

Screen Capture 16.5: One DAB will do ya'. Keep your eye on that Remote S Reg button as we'll be clicking on that puppy soon.

As expected, all is good in S-Register land. The AT+TOKDUMP command did its thing and I checked its work with a click on Display All:

```
**************************************************************************
    AT+TOKDUMP
    S00:0001
    S01:1DAB
**************************************************************************
```

Recall that we were able to name nodes using ASCII characters in the Rabbit Semiconductor ZigBee implementation. Well, we can do that with the ETRX2 as well. Just for grins let's use the command in Screen Capture 16.6 to name this node MOTHERSHIP. Remember STARCHILD? Well, if you purchased the Parliament album like I asked you to, you would know that STARCHILD was a resident of the MOTHERSHIP.

Screen Capture 16.6: As STARCHILD would say, "If you hear any noise, it's just me and the boys."

So far, so good. As we're passing through these AT commands, put them into the context of originating from the UART of your favorite microcontroller. I pulled a Display All and here's where we are:

```
AT+TOKDUMP
S00:0001
S01:1DAB
S02:3
S03:<hidden>
S04:MOTHERSHIP
```

That's enough custom stuff for the PAN for right now. Before we test our S-Register selections by establishing a new PAN, I scanned for pans with a click to the Scan For PANs button in Screen Capture 16.7.

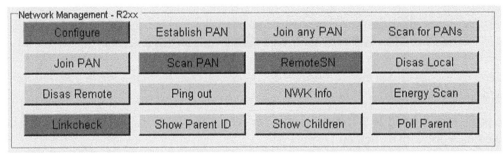

Screen Capture 16.7: The response to Scan for Pans command AT+N was +N=NoPAN.

Establish PAN is the next step in our journey. I clicked on that bugger and here's what was returned in the Telegesis Terminal window:

```
AT+EN
JPAN:11,1DAB
```

Obviously, AT+EN is the AT command to establish a PAN and the response contains our selected channel and our selected PAN ID. All of this time I had an ETRX2 scanning the 2.4-GHz channel space searching for a PAN to join. Once the 0x1DAB PAN came to life, this is what was displayed in the Telegesis Terminal window:

```
NEWNODE: 000D6F00000AD933
```

Now we can have some real fun. Check out the LED buttons in Screen Capture 16.8.

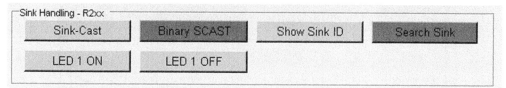

Screen Capture 16.8: All of the buttons you have seen me use in the Telegesis Terminal application are soft and programmable.

I know from the ETRX2 development kit documentation that the ETRX2 on the MCB has the pair of LED cathodes attached to bits 5 and 7 of the ETRX2 general-purpose I/O pins. Since there are only a couple of buttons and a couple of LEDs tied to the ETRX2's general-purpose I/O interface, I can turn the remote ETRX2's LEDs on and off by simply carpet bombing the ETRX2's general purpose I/O with 0x0000 and 0x00F0, respectively. That's exactly what I did in Screen Shots 16.9 and 16.10.

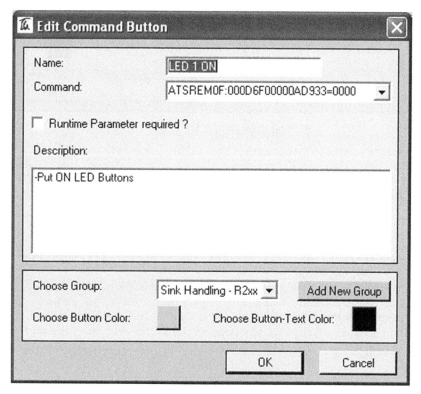

Screen Capture 16.9: Lights on...

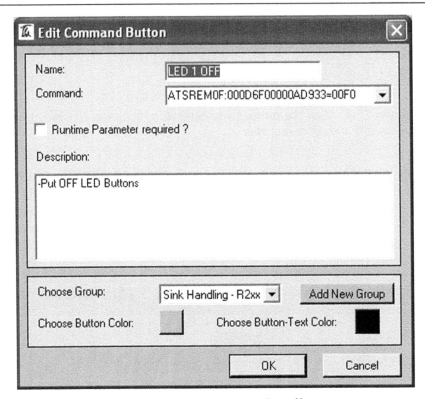

Screen Capture 16.10: Lights off...

The ATSREM command is the remote version of the ATS AT command. Trust me. When I click on the LED 1 ON button, the pair of LEDs on the remote ETRX2 node illuminate.

Suppose I want to read the status of the switch closures on the remote ETRX2 node. No worries. Here's the command:

```
********************************************************************************
ATSREM11:000D6F00000AD933?

S-Register S11 is the input buffer of the remote ETRX2's general purpose I/
O, while S-Register S0F is the output buffer.  The results of my depressing
various button combinations on the remote ETRX2's MCB are shown following
the initial read with no pushbutton activity:

ATSREM11:000D6F00000AD933?     // Both pushbuttons inactive
S11:0FF7
OK
ATSREM11:000D6F00000AD933?     //Both pushbuttons depressed
S11:0FF4
OK
ATSREM11:000D6F00000AD933?     //Pushbutton 1 depressed
S11:0FF6
```

```
OK
ATSREM11:000D6F00000AD933?      //Pushbutton 2 depressed
S11:0FF5
OK
```
****

Obviously, from the pushbutton trace results the MCB's pushbuttons are connected to the pair of least-significant bits of the ETRX2's general-purpose I/O. I created a button in the Sink Handling-R2xx group to obtain the status of the MCB's pushbuttons in Screen Capture 16.11.

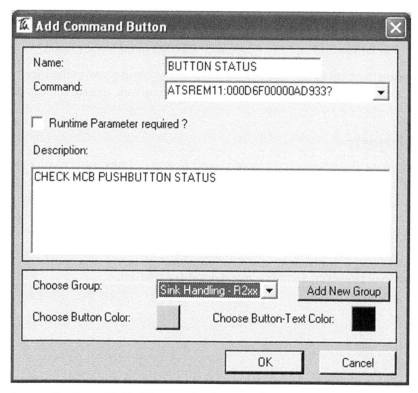

Screen Capture 16.11: The text in the Description window is displayed when the button is moused over.

Screen Capture 16.12 shows you what the new button looks like.

We spoke earlier about slowing down the traffic flowing in the default ETRX2 network. The ETRX2 contains eight individually programmable timer/counters that control such things as waking from sleep, when to poll for data, when to take an analog-to-digital converter reading and what power modes to assume depending on a timer.

The idea behind the ETRX2 transceiver is to allow the ZigBee network application programmer to use the well-knowN AT command-set template to configure the operational parameters of an ETRX2 node using a simple 3-wire serial port and the ETRX2's internal S-Registers.

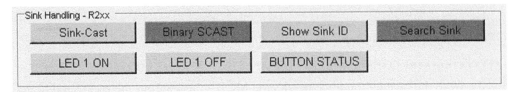

Screen Capture 16.12: The Telegesis Terminal application can be used to completely test all of your projects' AT commands before you hardcode them into your microcontroller-based ETRX2 application.

From an implementation standpoint, using the ETRX2 transceiver modules is a no-brainer. Just call Danny at Lemos and tell him how many ETRX2 modules you want.

If the built-in functionality of the ETRX2 modules doesn't suit your application's needs, you can use Ember's Insight development toolsuite to totally customize the operation of the ETRX2.

About Telegesis

My association with Telegesis is by way of Danny Lemos at Lemos International. Danny and the folks at Telegesis were also first-responders to my cry for help with ZigBee. In fact, many of the cutting-edge RF products, such as the ETRX2, that I have the opportunity to write about are products offered by Lemos International. Thanks, Danny, for your contribution to the content of this book.

I don't know about you, but singing "A Boy Named Sue" in front of a bunch of inmates could have consequences. However, Johnny Cash did it more than once.

Cypress MicroSystems's CapSense

Does PSoC mean anything to you? If PSoC is not in your vocabulary, here's a weather satellite view. PSoC is a product of Cypress MicroSystems and mixed signal is PSoC's claim to fame. A PSoC device incorporates internal analog and digital building blocks that are connected by firmware. The PSoC building blocks interact in input to output fashion using the PSoC's internal multiplexer and bus infrastructure. A PSoC application allows one to string together and interconnect comparators, timers, PWM modules, digital communication blocks (RS-232, IrDA, I²C, SPI, etc.), counters, analog-to-digital converters, digital buffers and digital inverters, which can generate and process real-world signals according to the firmware you feed into the PSoC's on-chip microcontroller. The latest addition to the PSoC mixed-signal array is the CapSense CSR (Capacitive Sensor Relaxation Oscillator) User Module.

Capacitance-based touch sensors are easily realized with mixed-signal PSoC devices. The Cypress PSoC engineers have put together a combination of PSoC hardware, practical capacitive touch sensor design guidelines and a capacitive sensor library complete with API calls that makes putting together a PSoC capacitive-based touch sensor design quick and easy. If you want to jettison the mechanical switches in your ZigBee or IEEE 802.15.4-based design, the CY3212-CapSense Training Board and the theory behind it will provide you with the knowledge needed to replace those mechanical switches with simple capacitive touch pads that can be etched directly on your ZigBee or IEEE 802.15.4-based device's printed circuit board. In addition, the PSoC device providing the touch pad service can focus its remaining digital and analog resources on other parts of the application that the touch pads are supporting.

Capacitive Sensing Basics

Sports-drink commercials bombard you with the thought that you need to drink their products to replace those electrolytes your body loses when you sweat. Those very same electrolytes you're running out of your pores are the reason why human fingers and capacitive touch pads work so well together. The electrolytes we carry around every day, coupled with our lossy dielectric-like skin, allow us to hold an electrical charge.

A typical capacitive touch sensor consists of an electrically isolated printed circuit pad sitting inside of a ground-plane area. Isolation of the printed circuit pad is provided by a 0.040-inch gap etched around a 0.40-inch pad. The result of this pad/ground plane arrangement is a simple capacitor. Unlike the enclosed parallel-plate capacitors you mount on your project boards, our CapSense-oriented capacitor plates are side by side, exposing some of the electrical field bands, which form energy bands between the capacitor's pair of open plates and ground. Because a finger is able to conduct and thus hold a charge, placing a finger in the proximity of our open-face pad/ground-plane capacitor increases the conductive surface of our simple open-faced printed circuit capacitor. The additional conductive surface supplied by the intruding finger adds additional charge storage capacity, which we can see as an increase of capacitance between the capacitive sensor and ground. Interestingly enough, the finger adding the capacitance does not have to be grounded. In fact, the finger's added capacitance can be sensed if the finger is grounded or floating. Sensing of the finger's capacitance is accomplished by the finger's charge participating in the electric fields spilling from around the pad/ground plane sensor. This interaction of fields implies that the finger doesn't have to physically touch the capacitive sensor to be sensed. The ability of the capacitive sensor to sense the finger without physical contact allows a thin nonconductive covering such as glass or lead-free plastic to be placed over the sensor.

CapSense Basics

All we need now is a way to sense the change in capacitance that occurs when a conductive object comes within proximity of our open-faced capacitive sensor. If we can sense a capacitance change in one capacitive sensor, there's no reason why we can't do the same for an array of capacitive sensors. However, if we decide to deploy an array of capacitive sensors, we must have a way to interrogate each of the sensors individually and report the sensor's status. That's where the PSoC comes in.

CapSense is implemented within a PSoC using a relaxation oscillator circuit consisting of a current source, a sensor capacitor, a comparator and a discharge switch, as shown in Figure 17.1.

The arrangement of analog and digital building blocks in Figure 17.1 is officially called a CSR User Module. The sensor capacitors you see depicted in Figure 17.1 are actually an array of capacitive touch sensors.

When selected by the analog mux, the capacitive touch sensor is attached to the CSR User Module and begins to be charged by the CSR User Module's DAC-controlled current source. Meanwhile, the PWM is set to its start value and the oscillator is enabled. The CSR User Module's comparator will trip and close the discharge switch when the voltage across the capacitive sensor reaches the comparator's preset threshold voltage. The charging ramp and the discharge cycle create a sawtooth waveform across the capacitive sensor. As the sensor charges towards the comparator threshold voltage, the comparator's output remains low. When the comparator threshold voltage is reached, the comparator output will go high.

By design, the comparator trip time and reset switch add a fixed delay of two system clocks. The output of the comparator is synchronized with the system clock to make sure that ample time is allowed to completely reset the charging voltage on the selected capacitive sensor. Each time the comparator threshold is reached and the comparator output changes state, a high-going pulse will appear at the output of the comparator. The CSR User Module's PWM is counting these comparator output pulses. While the PWM is counting comparator output pulses, the 16-bit counter gate is enabled, allowing the 16-bit counter to accumulate clocks

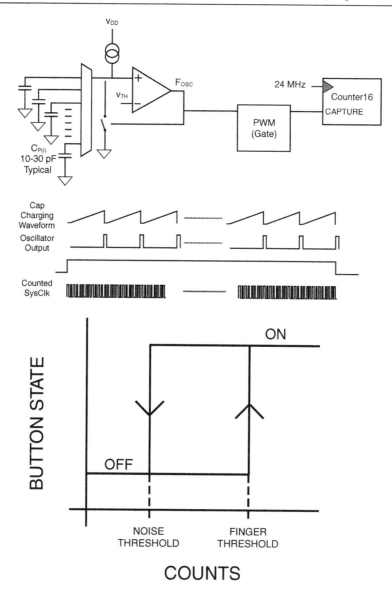

Figure 17.1: The sensor capacitors in this diagram are actually capacitive sensors that respond to the proximity of a conductive object such as a finger. The PSoC allows each sensor to be muxed into the CSR (Capacitive Sensor Relaxation Oscillator) circuitry.

from its 24-MHz signal source. The PWM will eventually count down to zero and disable the gate signal to the 16-bit counter. The raw count yielded by the 16-bit counter is then transferred to the CSR_iaSwResult integer array. The counting of pulses by the PWM module is associated with what Cypress calls the Period Method of measurement.

If you use your imagination to stretch and compress the sawtooth cycles you see in Figure 17.1, you'll come to the conclusion that the larger the sensor capacitance, the longer the ramp of the sawtooth cycle will be, which means more time between PWM counts, resulting in a longer 16-bit counter gate time and more counts accumulated by the 16-bit counter.

The determination as to whether a finger is influencing the sensor is based on the difference in baseline counts versus finger counts. Initially, baseline counts are accumulated at start-up time while there is nothing in the proximity of the capacitive sensors. Each sensor is scanned and its initial baseline value gets stored in the baseline array. The baseline is historically updated on every user-determined number of scans defined by the Update Rate parameter. Keeping track of the baseline insures that changes in the characteristics of the capacitive sensor due to environmental changes are accounted for.

Noise Threshold is defined as the number of counts below sensor turn-off. If the sensor's difference value is within the positive and negative limit of the Noise Threshold, the baseline value of that sensor is updated using an IIR filter. The IIR filter adds 75% of the previous baseline and 25% of the current raw data count to form the new baseline value. If the sensor's difference value is in the dirt as far as noise is concerned, the current raw data count becomes the baseline value. If the sensor isn't in the dirt and the difference value is not within the positive and negative limit of the Noise Threshold, the baseline value is not altered. The opposite of Noise Threshold is Finger Threshold, which is defined as the number of counts needed to consider the sensor as "ON". The combination of the Noise Threshold and Finger Threshold cut-off points create a welcomed hysteresis effect.

The baseline count is a 16-bit variable with a maximum value of 0xFFFF. The CSR User Module's baseline adjust routine multiplies the raw baseline count by four to maximize resolution. Thus, our raw finger count cannot exceed 0x3FFF counts. The PSoC documentation doesn't mention running the CSR User Module at any other system clock frequency except 24 MHz. So, our maximum sensor count time at 24 MHz is 42 nS (1/24 MHz) multiplied by 16383 (0x3FFF), which gives us 682 μs.

The idea behind the CSR User Module is very simple and it works very well. If you've been around as long as I have, you've built similar capacitive-controlled circuits using the famous LM555. If you're just a young pup and wouldn't know a 555 if it fell on you, it may interest you to know that you can build up an emulated LM555 using the elements of a PSoC. If you're interested in the LM555 in PSoC clothing, there is a PSoC application note that tells you all about it. However, we're not here to create PSoC 555 timers. So, let's take a look at the CapSense hardware implementation offered by Cypress.

CapSense Hardware

My CY3212-CapSense Training Board is shown in Photo 17.1.

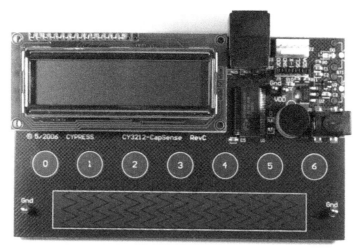

Photo 17.1: The CY3212-CapSense Training Board is very unassuming. There's just enough stuff here to sense your finger on the pads and slider. The CY3212-CapSense Training Board includes a CY8C21001 OCD part, which adds a bunch of advanced debugging clout to this unassuming development board.

As you can see in Photo 17.1, the CY3212-CapSense Training Board's capacitive sensor array consists of seven etched buttons and a slider, which is made up of 16 capacitive sensor segments. All of the CapSense work is done by a single CY8C21001 OCD (On-Chip Debugger) part. The CY8C21001 stands in place of the CY8C21434 PSoC that it emulates on the CY3212-CapSense Training Board. Using the CY8C21001 instead of the stock CY8C21434 PSoC allows the human PSoC CapSense designer to use the more powerful PSoC emulation and debugging tools for CapSense project development.

Speaking of programming PSoC devices, the CY3212-CapSense Training Board comes with a PSoC MiniProg module, whose likeness has been captured in a photograph in Photo 17.2.

Photo 17.2: Using this handy little device, I was able to program my CapSense code changes into the CY8C21001 and run them without having to remove the PSoC MiniProg module from the CY3212-CapSense Training Board ISSP socket.

The PSoC MiniProg is driven by a USB-enabled program called PSoC Programmer. PSoC Programmer has the ability to instruct the PSoC MiniProg to reset and power the CY8C21001 after programming, which is very handy when you're running through quick little code changes. The PSoC MiniProg worked so well for me that I not once pulled out my PSoC emulator. Another plus the PSoC MiniProg module provides is to eliminate the need for a wall wart.

The 16 × 2 LCD mounted on the CY3212-CapSense Training Board is used to visually indicate which capacitive sensor is excited. One of the coolest things you can do is slide your finger across the sensor while checking out the LCD. The default CapSense demo program not only indicates numerically where your finger is on the slider but it also tells you which direction you're moving it (right or left). For those of you that are musically inclined, a speaker is also part of the CY3212-CapSense Training Board. I've pulled up close on the business end of the CY3212-CapSense Training Board for you in Photo 17.3.

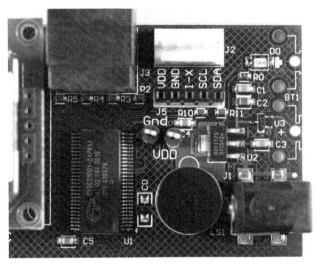

Photo 17.3: Another advantage of using PSoC devices is the reduced number of popcorn and glue parts needed by the final PSoC hardware. Most of the stuff you see in this picture would be eliminated if this were to become a commercial product. Note the hatch groundplane, which is recommended and specified in the CapSense design documentation.

CapSense Logic

I'm sure interactive books are on their way. However, they ain't here yet. So, showing you what happens on the LCD every time I touch a sensor isn't going to happen right now. Instead, I'll physically remove the LCD, make a slight PSoC hardware change and gain access to the CY8C21001's CapSense results by way of good old RS-232.

Before we make our modifications, let's look at the PSoC Designer view of the original CSR User Module hook-up that you get when you run the LCD demo that comes loaded on the CY8C21001. Shift your attention to Figure 17.2 for a moment.

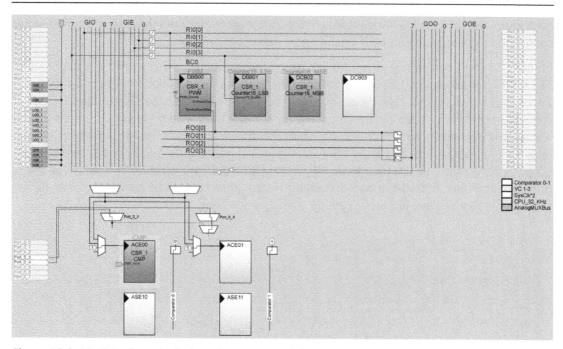

Figure 17.2: PSoC Designer includes a CSR User Module wizard that is used to assign the CSR_1 sensor inputs to their respective PSoC general-purpose I/O pins. Note that the LCD is assigned exclusively to Port 2. The CSR User Module is cast in stone and will always populate the PSoC resources and make the connections you see in this figure.

The little tag labeled with a "1" feeding the ACE00 comparator module's clock input is the PSoC's 24-MHz system clock. Although the PSoC Designer view doesn't show it in detail, the comparator's input is actually muxed to the capacitive sensor buttons that are attached to the PSoC's Port 1 and Port 3 general-purpose I/O pins using an internal AnalogMuxBus. This internal AnalogMuxBus is also used to connect the CSR User Module's DAC current source and reset switch.

The comparator's output, which is also not obvious to the most casual observer, is routed internally to GOO4 (GlobalOutputOdd4). An interconnect resource takes the GOO4 signal to GIO4 (GLobalInputOdd4), which is connected to RIO[0]. The PSoC PWM module gets its input from RIO[0], whose signal path we just traced back to the comparator output.

The PWM module output feeding ROO[3] is passed to the NOT-input side of the [~A AND B] gate fed by ROO[3]. The remaining PWM module output takes the ROO[0] rail to the B side of the ROO[3] [~A AND B] gate. The output of the [~A AND B] gate feeds GOO7 and interconnects to GIO7, which feeds into RIO[3] and the 16-bit counter module's input.

Figure 17.2's hardware layout uses an associated PSoC C program to put the number of the sensor touched on the LCD. I want to show you some count values associated with the sensor activations and the best way to do that is to feed the count values out through a standard serial connection to a terminal emulator running on a personal computer.

Figure 17.3 is a PSoC visual of my version of the touchy-feely CSR User Module application. The CSR User Module is laid in concrete and there are no changes that I could make there. I need to get at the CY8C21001's general-purpose I/O pins that are attached to the LCD, as they are the only way I can get signals out of the CY8C21001 without having to micro-solder wirewrap wire directly to the CY8C21001 pins. (I think you'll see what I mean when you look back at Photo 17.3.) So, using PSoC Designer I eliminated the LCD connections on the CY8C21001's Port 2 general-purpose I/O pins and pulled the physical LCD from its 14-pin SIP socket on the CY3212-CapSense Training Board. I can stick some header pins into the LCD SIP socket and use jumpers to get the signals I need out.

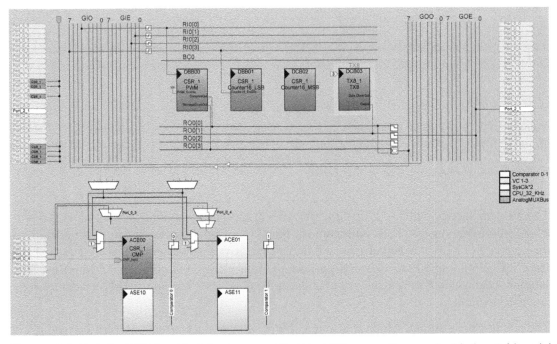

Figure 17.3: I left the CSR User Module alone, scuttled the LCD connections and added an 8-bit serial transmitter user module. I used the PSoC Designer resources to assign a clock to my new transmitter and make the connection from the transmitter output to a PSoC general-purpose I/O pin.

The PSoC module library includes separate serial transmitter and serial receiver user modules. I really don't need to receive any data here. So, the PSoC 8-bit serial transmitter module will do just fine. After loading the CSR User Module into the configuration, the architecture of the CY8C21434 that is being emulated by the CY8C21001 leaves me with a single digital module (DCB03), which I can put to good use as an 8-bit serial transmitter. To use the DCB03 as an 8-bit serial transmitter, all I have to do is set the baud clock and connect the output of the 8-bit serial transmitter to a CY8C21001 general-purpose I/O pin left open by the removal of the LCD. Setting the baud clock is an easy task, as the PSoC requires the 8-bit serial transmitter be clocked at eight times the bit rate. I want to run the 8-bit transmitter at 115200 bps. So, with a 24-MHz system clock, I need to divide the incoming system clock by 26 to achieve the 8× bit clock. Performing the system clock division is an easy task, as all I have to do is specify the system clock source, specify the clock output portal and dial in a divisor of 26 for the output portal in the PSoC Designer Global Resources window. Once I have the bit-rate parameters set, all that's left to do is connect the serial transmitter module's output to one of the CY8C21001's Port 2 general-purpose I/O pins. The PSoC CY8C21001 doesn't have the resources available to do the digital-to-RS-232 voltage-level conversion. That's OK, as I have literally dozens of prototypes containing the necessary TTL-to-RS-232 conversion circuitry that I can tap the serial transmitter's output into.

The capacitive sensors of the CY3212-CapSense Training Board are in the open and are not covered with an overlay. The absence of an overlay makes the sensors very sensitive. Thus, the scan speed of the CY3212-CapSense Training Board sensors is set at the minimum of three, which scans for a single oscillator period. Three is the minimum for a single oscillator period as two oscillator periods are added to compensate for the PWM interrupt overhead. The DAC current also has to be adjusted for the conditions, as the number of raw counts is inversely proportional to the amount of DAC current supplied to the sensor. In one of the CapSense application notes, the Cypress CSR User Module application engineer who authored the app note points out that capacitive touch sensing is an art. With that in mind, the calibration extents of the CY3212-CapSense Training Board sensors have been determined to lie within a range of 290 to 310 raw counts with no fingers present. A simple loop increments or decrements the DAC-controlled current source value for each of the seven individual sensors using the CSR_StartScan and CSR_SetDacCurrent API calls until each sensor raw count falls into the calibration extents that were specified. I read in another CapSense document that much of what one has to deal with when designing capacitive touch-sensing applications is left up to experimentation. I'm sure that's where the 290–310 numbers came from, as there is absolutely no technical or mathematical explanation of how these numbers were chosen in the CapSense documentation. Personally, I like to use SWAG design methods every now and then. Right now, the Finger Threshold is set for 100 and the Noise Threshold value is set to 40 within PSoC Designer. Given the aforementioned values were most likely derived from the physics of the CY3212-CapSense Training Board, I'll leave them be. I'll also let the current BaselineUpdateRate value of 40 stand.

Once all of the sensors are calibrated, we can call upon the CSR_bUpdateBaseline API for the interrogation of the sensors one by one. Because the CSR User Module is multiplexing the sensors into a common CSR User Module configuration, the DAC current calibration value for each particular sensor is loaded before the selected sensor is scanned.

Here's what goes on inside the CSR_bUpdateBaseline API call. The CSR_bUpdateBaseline API call interprets the raw sensor data and is responsible for many of the numbers I generated with a touch of the CY3212-CapSense Training Board's 0 sensor in Figure 17.4.

Sensor	waSnsResult	baSnsOnMask[0]	waSnsDiff[0]	bBaselineUpdateTimer	waSnsBaseline/4
0	01B4	0001	0088	0028	012C
0	01B4	0001	0088	0027	012C
0	01C2	0001	0096	0026	012C
0	01C1	0001	0095	0025	012C
0	01C7	0001	009B	0024	012C
0	01C3	0001	0097	0023	012C
0	01C9	0001	009D	0022	012C
0	01D0	0001	00A4	0021	012C
0	01CE	0001	00A2	0020	012C

Figure 17.4: The sensor with a finger on it is indicated by the bit mask in the baSnsOnMask column. Raw counts are tallied in waSnsResult and the baseline counts used in the difference calculation are found in the waSnsBaseline/4 column. The waSnsDiff[0] value is used by the CSR_bUpdateBaseline API call to make an "ON/OFF" decision.

A difference value, which is stored in waSnsDiff[0] for sensor 0, is calculated by subtracting the previous baseline value (waSnsBaseline/4) from the raw data (waSnsResult) obtained from sensor 0 (zero). Note that in Figure 17.4 the bBaselineUpdateTimer value begins with 0x28 and every time the CSR_bUpdateBaseline API call returns a positive result (a sensor is "ON") the bBaselineUpdateTimer value is decremented. When the bBaselineUpdateTimer value exhausts itself, the CSR_bUpdateBaseline API is eligible to perform an update on the sensor's baseline value depending on the Noise Threshold value, as I described earlier. For us, the most important thing the CSR_bUpdateBaseline API call does is determine if the sensor is "ON" or "OFF". The determination of the sensor status is stored in the baSnsOnMask[0] array as a "1" for "ON" and a "0" for "OFF".

I've provided the listing (Listing 17.1) that generated the numbers you see in Figure 17.4.

Listing 7.1

```
***********************************************************************
//-----------------------------------------------------------------
//   For Board:  CY3212-CapSense RevC
//   Chip:       CY8C21434-24LFXI
//
//   Project Settings (in the Device Editor):
//
//      Global Resources:
//          Power Setting [Vcc/SysClk freq]: 5.0V / 24MHz
//          CPU_Clock:    SysClk/1
//
//      User Module Parameters:
//          Method: Period
//          FingerThreshold: 100
//          NoiseThreshold: 40
//          BaselineUpdateRate: 40
//          ESDDebounce: Disabled
```

```
//-------------------------------------------------------------------
// bit masks defined for each button
#define BUTTON_0 0x01
#define BUTTON_1 0x02
#define BUTTON_2 0x04
#define BUTTON_3 0x08
#define BUTTON_4 0x10
#define BUTTON_5 0x20
#define BUTTON_6 0x40
// define the number of switches in this system
#define NUM_SWITCHES 7
// the followin defines the ranges for the raw counts
// during the calibration routines
#define DAC_MAX_RAW_COUNT   310
#define DAC_MIN_RAW_COUNT   290

void CalibrateSwitches(void);
// Array that holds the individual DAC settings for each switch
BYTE bDACcurrent[NUM_SWITCHES];
WORD i,x;
//-------------------------------------------------------------------
void main()
{
    CSR_1_Start();   // call CSR Start, does all the internal CSR connections
    M8C_EnableGInt; // Enable global interrupts
    TX8_1_Start(TX8_1_PARITY_NONE);
    CalibrateSwitches();

    while(1) // start the main loop
    {
        // scan each switch individually
        for (i=0; i<NUM_SWITCHES; i++)
        {
            CSR_1_SetDacCurrent(bDACcurrent[i],0); // Sets DAC current
            CSR_1_SetScanSpeed(3); // use a scanspeed of three (no overlay so
buttons are very sensitive)
            CSR_1_StartScan(i,1,0); // Scan one Switch (i)
            while(!(CSR_1_GetScanStatus() & CSR_1_SCAN_SET_COMPLETE));
// wait until the scan is complete
        }

        // call the update baseline function which does several functions, the
two main
        // functions it provides are:
        //  1) updates each switches baseline (if necessary and according to
the rate
        //     set by the updateBaseline value
        //  2) determine the button status (on/off) and places this value
        //     in CSR_1_baSnsOnMask[]
        //  if any of the buttons are determined to be "ON" then the function
will
        //   return a non-zero value
```

```
            if(CSR_1_bUpdateBaseline(0))
            {
                // check to see if BUTTON 0 is on or off
                if ((CSR_1_baSnsOnMask[0]) & BUTTON_0)
                {
                 TX8_1_PutChar('0');
                 TX8_1_PutChar(',');
                 TX8_1_PutSHexInt(CSR_1_waSnsResult[0]);
                 TX8_1_PutChar(',');
                 TX8_1_PutSHexInt(CSR_1_baSnsOnMask[0]);
                 TX8_1_PutChar(',');
                 TX8_1_PutSHexInt(CSR_1_waSnsDiff[0]);
                 TX8_1_PutChar(',');
                 TX8_1_PutSHexInt(CSR_1_bBaselineUpdateTimer);
                 TX8_1_PutChar(',');
                 TX8_1_PutSHexInt(CSR_1_waSnsBaseline[0]/4);
                 TX8_1_PutCRLF();
                }
                // check to see if BUTTON 1 is on or off
                if ((CSR_1_baSnsOnMask[0]) & BUTTON_1)
                {
                 TX8_1_PutChar('1');
                 TX8_1_PutChar(',');
                 TX8_1_PutSHexInt(CSR_1_waSnsResult[1]);
                 TX8_1_PutChar(',');
                 TX8_1_PutSHexInt(CSR_1_baSnsOnMask[0]);
                 TX8_1_PutChar(',');
                 TX8_1_PutSHexInt(CSR_1_waSnsDiff[1]);
                 TX8_1_PutChar(',');
                 TX8_1_PutSHexInt(CSR_1_bBaselineUpdateTimer);
                 TX8_1_PutChar(',');
                 TX8_1_PutSHexInt(CSR_1_waSnsBaseline[1]/4);
                 TX8_1_PutCRLF();
                }
////////////////////////////////////////////////////////////////////////
//
// Function: CalibrateSwitches()
//
// Description:
//    This function will automatically determine the appropriate
// DAC values for each switch in the system.  The values will be
// stored in the variable - bDACcurrent[].  The function will set
// the DAC value depending on the values set by the following lines:
//
// #define DAC_MAX_RAW_COUNT    310
// #define DAC_MIN_RAW_COUNT    290
//
// These lines are located at the top of the program and give a range
// for the raw CSR count values to be in to set the DAC value.  This
// function will iterate through each switch.  It will do a single
// scan and compare the raw count values with the range defined above.
```

```
// If the count values for that particular switch are outside of the
// range, the function will change the DAC value appropriately and do
// a rescan.  This process is repeated until the raw counts for each
// particular switch are within the specified range.
//
void CalibrateSwitches(void)
{
    BYTE bFlag = 1;

    for(i=0; i<NUM_SWITCHES; i++) // iterate through all the switches in the
system
    {
        // calibrate DAC setting for a switch
        //----------------------------------------------------
        bFlag = 1; // reset the flag to a true state
        CSR_1_SetScanSpeed(3); // start with a scanspeed of three to get ONE
oscillator cycle or period
        bDACcurrent[i] = 20; // a DAC current of 20 will start us in range

        // do one scan and compare the raw count value against
        // the defined range.
        // - If the counts are BELOW the set range
        // DECREASE the DAC setting to INCREASE the counts
        // - If the counts are ABOVE the set range
        // INCREASE the DAC setting to DECREASE the counts
        do
        {
            // set the DAC current
            CSR_1_SetDacCurrent(bDACcurrent[i], 0);
            // scan just the one button
            CSR_1_StartScan(i,1,0);
            while(!(CSR_1_GetScanStatus() & CSR_1_SCAN_SET_COMPLETE));

            // check to see if the counts are in range
            if (CSR_1_waSnsResult[i] < DAC_MIN_RAW_COUNT)
            {
                bDACcurrent[i]--; // counts are BELOW the range so DECREASE the
DAC setting
            }
            else if (CSR_1_waSnsResult[i] > DAC_MAX_RAW_COUNT)
            {
                bDACcurrent[i]++; // counts are ABOVE the range so INCREASE the
DAC setting
            }
            else
            {
                bFlag = 0; // the counts are in range so exit the do loop
            }
        }while (bFlag); //if the flag is still 1 then go back and rescan the
switch
    }
}
```
****

323

It was very refreshing to discover that the CapSense application engineers admitted that you will need more than the math and theory to build up a good CapSense implementation. However, the CapSense documentation and application notes provide more than enough of the engineering stuff you'll need to mix with your experimentation. The PSoC takes most of the behind-the-scenes complication out of putting an embedded capacitive touch application together. I've included this informational discussion centered on the CapSense technology because it can be directly applied to a ZigBee or IEEE 802.15.4 application.

About Cypress MicroSystems

If you haven't experienced PsoC, you owe it to yourself to get your hands on a PSoC development kit. PSoC devices are geared towards low-power operation and fit perfectly into sensor-based designs.

The Final Word

About halfway through the writing of this book, the ZigBee Alliance decided that they wanted to field a new ZigBee Specification. Oh, Boy! I figured all of my work up to that point was null and void. To make matters worse, the ZigBee folks weren't going to release the new spec to the general pubic until after this book was published.

I'm just a poorboy from Tennessee. So, I couldn't cough up enough money to join the ZigBee Alliance. However, I ain't too proud to beg (Temptations 1966 / Rolling Stones 1974). Governments around the world should consider looking into the ZigBee Alliance's security practices as I could not get one drip of advance information from any of the ZigBee Alliance members.

I've learned over the years that one must find an outlet for stress created by situations like this. So, I loaded up a few ammo cans, packed my heat (I prefer carrying GLOCKs) and jumped into the Jeep. My friend, Steve Kennedy, who happens to own The Gunsite, my favorite pistol range, welcomed me into his place and bent his ear to my dilemma. After listening to me whine, Steve just shook his head and handed me a bunch of targets.

After the smoke cleared, I did some web surfing and noticed that not one of the 802.15.4 radio manufacturers was changing their radios to meet the new spec. I also didn't see a rush to replace or update any of the development kit hardware I had already examined and written about. In addition, there wasn't any buzz about changes to the 802.15.4 2003 standard, as the 802.15.4 wizards didn't even get out of their South Beach hammocks. As the days passed, and more shots were fired at Steve's place, it became apparent that I should move on with this book and worry about the new ZigBee standard only if I had to. After all, how can I write about something that I can't obtain any information about? That's like shooting in total darkness.

A few weeks later, the ZigBee folks had become so proud of their new baby that they decided to release the new ZigBee 2006 specification to the general public. My Mom always says that worrying about things is a waste of time as you have no control of the natural events of your future. Again, she was right. The 2006 ZigBee specification lays out some really slick stuff. However, everything you've learned about the structure of ZigBee and the operations performed by 802.15.4 by reading this book can still be applied to the new 2006 ZigBee document. There's rumor that a 2007 ZigBee specification is already in the works. So what? We're covered.

Index

Printed and bound by CPI Group (UK) Ltd, Croydon, CR0 4YY

03/10/2024

01040319-0003